FOUR HEDGES
A Gardener's Chronicle

Written and engraved by

Clare Leighton

LITTLE TOLLER BOOKS
an imprint of THE DOVECOTE PRESS

This paperback edition published in 2010 by
Little Toller Books
Stanbridge, Wimborne Minster, Dorset BH21 4JD
First published in 1935

ISBN 978-0-9562545-3-5

Typeset in Monotype Sabon by Little Toller Books
Printed in Spain by GraphyCems, Navarra

All papers used by Little Toller Books and the Dovecote Press
are natural, recyclable products made from
wood grown in sustainable, well-managed forests

A CIP catalogue record for this book is available
from the British Library

3 5 7 9 8 6 4

CONTENTS

FOREWORD   11
David Leighton

INTRODUCTION   13
Carol Klein

April   21

May   34

June   48

July   62

August   75

September   86

October   99

November   112

December   125

January   136

February   147

March   158

ENGRAVINGS ON WOOD

Cowslips	21
Grape Hyacinths	23
Hyacinth Bud	25
Anemone Pulsatilla	26
Daffodils in Bud	29
Horse Chestnut Bud	30
Blackbird on Nest	31
Auricula	33
Cherry Blossom	34
Crown Imperial	37
Columbine	38
Weeds	41
Wild Arum	42
Oriental Poppies	45
Deserted Thrush's Nest	47
Nest of Hedge Sparrows	48
Raspberries	50
Fledgling Wrens	53
Rose	54
Picking Strawberries	57
Peas	58
Mullein Caterpillar	60
Moth	61
Broad Beans	62

Sharpening the Scythe   64

Scything   67

Bindweed   69

Lily   71

Blackcurrants   72

Snail   73

Frog   74

Gooseberries   75

Poppy Heads   77

Plums   79

Snapdragon   81

Tortoise   82

Sunflowers   83

Grasshopper   85

Blackberries   86

Pears   89

A Lapful of Windfalls   91

Vegetable Marrows   92

Japanese Anemone   94

Autumn Crocuses   95

Hedgehog   96

Swallows   98

Digging Potatoes   99

Sloe   101

Berries   105

Hazel Nuts   108

Slug   110

Horse Chestnut   111

Transplanting Walnut Tree   112

Last Leaf on the Cherry   115

Tits and Sunflower Seeds   117
Bonfire   118
Leeks in Flower   120
Planting Trees   123
Brussels Sprouts   124
Maple Seeds   125
Sweeping up Leaves   127
Earthworms   129
Fungus   130
Hazel Catkins   133
Yellow Jasmine   134
Axe and Block   136
Tits on Hopper   139
Snowdrops   141
Winter Aconites   142
Iris Stylosa and Christmas Roses   144
Garden in Snow   147
Scylla   149
Coltsfoot   150
Pyrus Japonica   153
Stinking Hellebore   155
Ivy Berries   157
Cyclamen   158
Green Woodpecker   160
Hepaticas   162
American Currant   165
Warm Weather Coming   166
Violets   167
Tulip Kaufmanniana   169
Daffodil Shoot   170

FOREWORD

David Leighton

CLARE ADOPTED WOOD-ENGRAVING as her favourite medium as early as her student days, and produced over 250 catalogued engravings before *Four Hedges* appeared in 1935. These earlier works include many book illustrations as well as six large engravings from her stay at a Canadian lumber camp and the twelve full-page illustrations to her own major book *The Farmer's Year*.

Notwithstanding Clare's later success in America, where she was elected a member of the National Academy and awarded an honorary doctorate, *Four Hedges* may well represent the happiest time of her long life: she was in her own home, with her own purpose-built studio, and sharing the creation of a garden with her long-term partner, the political journalist Henry Noël Brailsford.

I am delighted that Little Toller Books has decided to produce a new edition of *Four Hedges*. It ran to several editions on first appearance, and I am confident that a new generation of readers will share that delight.

INTRODUCTION

Carol Klein

I WONDER IF, when Clare Leighton and her partner Noel Brailsford took on their garden in the early thirties, she had any idea of the fact that recording its progress, documenting its changes and passing on her observations of what would happen within its 'Four Hedges', would occupy as much of her time as reorganising its borders and scything its grass.

It was to become a major part of her life. Though it must have been hard to follow the digging by drawing, to form sentences and paragraphs about what had just been finished when she came in from a hard day's work, her hands blue with cold and covered in mud, or her dress wet with sweat from scything the orchard.

Not only are her observations on how things happen in the garden generic (anything closely observed and accurately reported is bound to have pertinence), but they are tempered with an enormous respect for all she sees.

She acknowledges the magic and dispels the mystique. There is no argument about what she describes, you trust her implicitly. Nowhere does she profess to 'know' about gardening. At every step she acknowledges that she is finding out and she fills you with the desire to do the same, to jump in and partake.

Fellow gardeners will recognise so many of the sentiments she expresses and identify with the activities she describes. Who hasn't drooled over bulb catalogues and then imagined, having placed an order, that the job is done and the bulbs will plant themselves.

The time and the context of the world that Clare describes have gone forever. On one level her book is a historical document, a record of the practices in gardening and farming in a Chiltern village in the early thirties, and yet her subject matter is timeless, an enduring record, as urgent as the work that needs to be done today.

I have admired Clare Leighton's woodcuts for many years. A couple of reproductions have been amongst my most treasured possessions, probably removed from one of her books and pasted onto card by some unscrupulous bookseller, they have delighted me nonetheless, pinned up in easy view on a year planner and removed and repinned when December gives way to January. Though their style is easy to date, the cut of a jacket, the sweep of a line, there is a timelessness about them. These men planting a tree, one with a long apron, both with flat caps, neither with gloves, may be stereotypical of 1930s rural workers but they have the essence of men planting trees that stretches back to the first time a tree was planted and all the subsequent times this noble activity has been undertaken and will be again. The landscape that encompasses them is ageless too, thin soil stretched over the spine of a sleeping giantess, a knoll of trees iconic on the close horizon, immediate yet eternal under a windswept sky.

I had no idea who had made these uncompromising images, permeated by the earth. I did not even know that it was a woman who had created this magic, though there is a huge empathy and admiration for those she portrays, human, animal and plant, that leads you to feel a loving female hand was responsible. There is much of the mother, the nurturer, in her writing. Images in word and picture are often cradled.

I had never read her writing until I was asked to write this introduction, but having read just a few pages I cried. Don't get the wrong impression. This is not sentimental or slushy stuff, it has to do with the power of the goddess. Every syllable, every scoop of the chisel, exudes Mother Earth.

The story of Clare's garden has no beginning, middle or end. That is what separates it from so much garden-writing, where ultimately most titles have conclusions. That is not to say that there is anything unformed, random or ill-considered about it: no room for accidents or eccentricities, only a massive respect for nature as it happens and an acknowledgement of its order, pattern and predestination. There is the strong, organic form in her imagery that permeates all natural processes. Art, like gardening, is an artificial activity, but at its best it can extract the essence from whatever it seeks to portray.

There must be all sorts of reasons for writing a book. *Four Hedges* is a labour of love. It feels as though Clare Leighton had no other option than to write and illustrate it. She had to record the arrival of the swallows, she had to describe the monochrome simplicity of a new dawn and within it the clarity of the song of the first thrush and she had to share its secrecy with her reader. It becomes a special secret between the two of us. No matter how many people have read and will read this book everyone will feel that she is speaking directly to them, showing them the sweep of the reaper's arm, helping them feel the weight of the apples, how the grass smelt, how cleggy the earth felt. There are moments of astonishment, deep joy and humour. Who wouldn't identify with pulling off the gardening gloves, the better to get to grips with the soil, to feel it and love it, abandoning not just gloves but all the inhibition and restraint that deny the very visceral experience of gardening. Without being subjected to any deliberate manifesto we are persuaded, we want to join the cause. We yearn to find in our own garden the wonder Claire elevates from the familiar.

Clare Leighton's melodious writing, coupled with the percussion of her uncompromising woodcuts, strikes a chord which resonates not just immediately but for ever. Phrases and images fill you with delight and months, years later, stick with you. Though I have only recently been lucky enough to read her writing, I know that it will stay always. Such is the joy you feel when experiencing her writing you immediately think you must persuade others unacquainted with her work to enrich their lives by reading this book and to do it straight away.

As you read each word, as you study each line, you smile. Yes, you want to say, yes, yes, you want to shout, that's it! That is how the blackbird sits on her nest as if to make sure that no breath of air, not even a fine feather's breadth, could squeeze between the curved wall of her nest and know the secret of the precious eggs that lie beneath her warm and patient body.

Four Hedges is documentation of the highest order. Clare's power of observation is razor sharp, direct and unremitting – but it is much more. It is tempered always with love and an enormous joy in being able to experience whatever she is telling us about.

This is the most honest writing I have ever read. There is no sophistry, no attempt to say something clever. No doubt there was frustration, no doubt difficulties, rewriting and editing can be irksome even debilitating, but if so it doesn't show. There is no loss of energy and momentum. Even when the subject she is describing is subtle, her words respect its nuances and recognise its life force. The writing and the images have an immediacy that is as fresh as the first primrose. Both Clare's parents were accomplished writers and no doubt the rigours of writing were part of her consciousness long before she herself moved into authorship.

Making a woodcut defies shilly-shallying, you can't fudge. In painting (especially in oils) and drawing there is room for manoeuvre, the opportunity to change your mind. Not when you are engraving. Commitment is essential and Clare's writing echoes that commitment. It is as incisive and immediate as her images, so fresh you feel it was you who crouched down to come face to face with a frog, you who heard the ironstone dragged along the blade of the scythe with long, slow strokes and felt the edge of the blade, sharp and shiny in the hot summer sun.

Her image of a vegetable marrow tells you everything about the life of the plant and so much more, the swelling fruit lurking underneath the vast leaves, the flowers newly opened, as ephemeral as any flower could be and yet for their moment as strong and substantial as the leaves that surround them.

How could she capture this process, this life, in a medium that entails the accumulation of cut after cut after cut, of incredible patience and hour upon hour of painstaking work. Her confidence and self belief shine through and it rubs off on the reader. But her art is artless.

Four Hedges is not an instruction manual on how to garden. Nobody will turn to it to find out when to sow their broad beans or how widely to space their potatoes. Clare was learning as she went in the only real way any gardener can – by experiment and trial and error, but she did so with a fervour and determination that inspires the faint-hearted and galvanises the most hesitant into feverish activity. The great gift of her writing and illustration is its ability to lift us all into another domain, one full of wonderment and magic. Her work reminds us of our place in the overall picture, at one with the earth and full of wonder and joy to be born of it and to engage with it.

Carol Klein
North Devon, 2010

To my Companion
within
the Four Hedges

APRIL

OURS IS AN ORDINARY GARDEN. It is perched on a slope of the Chiltern Hills, exposed to every wind that blows. Its soil is chalk; its flower beds are pale grey. Dig into it just one spit, and you reach, as it were, a solid cement foundation. One might be hacking at the white cliffs of Dover. Only when it is wet from heavy rain does our soil darken and look normal. It is a new garden. In it there are none of the great trees that spread their shade over stretches of lawn, none of the mellow, age-silvered bricks that shelter a walled-in fruit garden, not a hint of a crazy paving patterned with moss, or a sundial with edges blunted by time. Four years ago it was rough meadowland, housing only larks and field mice. But a little over a hundred years ago it was common land which the rector of the day appropriated under an Enclosure Act. Grasses covered it throughout the year, changing in colour and height with the seasons as they waved and rustled unnoticed. Spring visited it in clumps of cowslips; wild roses opened full to the June sun. Its hedges yielded a rich harvest at the fall of the year, when spiders swung their hammocks from blackberry to sloe. In its bounding hawthorns birds mated and nested and fed their young, unchecked by the abrupt movements of men. Cuckoo-pints pierced the darkness of hedge-bottoms, and over the hedges themselves bindweed and bryony grew in twisting tangles.

We saw it first one shining day of May. We had walked among the hills, looking from meadow to meadow, balancing view against view, earth against earth. For we knew that were we to slip down to the plain, our earth would be rich and dark and our flowers would be sheltered and flourish. But two things made us buy this land: we could not withstand the appealing beauty of the clumps of cowslips nor resist the hawthorn hedges that bounded the grassland. The hedges foamed that day with bloom, low and thick where they had been layered, tall where hawthorn trees had been left standing. And at one end, where the hedge met the lane, stood a horse-chestnut, its blossoms lit.

So through four years around the nucleus of clumps of cowslips we have tamed and enriched this half acre of grassland, bending it to our will, fighting its stubbornness. Winds of unimagined force have battled against us, bursting through the gaps between our hills, sweeping unhindered across the midland plain, beating upon us through the pass in the east; wireworm and leather-jacket have riddled our uncultivated earth; where we have been lenient to flowering weed, it has repaid our kindness by spreading a hundredfold in soil that we have fed and turned.

The first thing we did was to plan our garden. We wanted good shapes. Noel, who was especially excited over it, spent days with pencil and paper, deciding where we should have lawn and flower bed and orchard, how many trees we should plant, what proportion of the garden should be given over to vegetables. As our bodies have skeletons so should our garden have bony structure. It is only upon a firm foundation that the irregular growth of plant and tree can best clothe and deck the garden. Frilled edge of flower contrasts with severe edge of design. In our garden we decided that we would have no sentimentality, no wilfully irregular edges to ornamental ponds, no badly-sculptured garden figures, no timidity showing itself in an escape into false "mossy bits", or an aping of the old-fashioned. This sentimentality of bad design has no historic heritage. It was by reason of its severe underlying structure that the really old garden scored its success; for what could be more rigid than the walled garden, with the tight symmetry of espalier fruit trees, ruler-straight lines of vegetable patch and unswerving order of box edging? We struggled with our shapes. The half acre turned out to be an imperfect rectangle and we had difficulty in compromising with the edges of flower beds and the lines of beech hedges. I did not myself mind greatly, but Noel, with his almost mathematical attitude towards the garden, was badly worried. Patterns were spun with trees, sentinel poplars stood at points of emphasis in the garden. The orchard was planted, a grass walk down its centre that we might wander between the trees in blossom time or fruiting time and see them in their varying clothing. Beneath the fruit trees we flung fans of daffodils. Around the garden we planned a grass path where we might walk quietly; but shade will not come for years.

And all the time the wind and the chalk battled with us, and the garden tugged back to the wild.

Now, in April, we walk up the lane to our gate. We have been away from the garden for a long time and we fear lest we have missed this year's spring. Shall we find faded daffodils in the orchard? Will the plum trees have scattered their blossom to the ground, or hyacinths have rusted? My year is out of focus, for I have returned from Corsica, where rose and iris bloomed in the hot damp air of March. So deeply stamped upon us is the ordered procession of the year that it is a wrench when it is dislocated, and we see the rose before the daffodil.

But when we enter our garden we find that our fears are groundless. Here is no spent spring, no tired blossoming. Our cold soil has coaxed no rush of flowers to bloom. Piercing grey winds shriek, and the birds are silent. In the orchard the clustering small shoots of daffodils stand like green druidical stones. The flowers of the hyacinths are clenched fists of bud. Everything looks strangely stark to me, with my mind still coloured by stretches of Mediterranean blossom and warm with the glow of the Corsican sun. But once recovered from the first impact, I seem to see with clearer eyes our cold northern light over everything and the beauty of the young growth of my plants. The drama of the

year is late in starting and I am in time for the first act.

We take our first walk round the garden, intimately noticing the many changes. Sadly we realise that we have missed the overture to this year, the snowdrops and crocuses; they hang colourless heads with swelling seed pod, or lie wind-dashed with their faces to the ground. In this late spring we are in time for the almond blossom; it is just opening its tight pink buds. The rhubarb is still covered against the frosts. Over the fence at the foot of the orchard our neighbour's old goat stands as she has stood for four years. As we go towards her, we frighten the timid green woodpecker from the tree stump at which he pecks; he rushes off with a disquieted chuckle. Clumps of green have appeared in the big perennial bed; thankful for the lateness of this season, we realise that we can still dig. With joy I find the shoots of my cherished yellow scabious plants; I had tended the seedlings all last summer but feared that they had died in the first frosts. There seems to be special pleasure in tending and strengthening sickly plants. I remember a fight we once had for some diseased Canterbury bells. The seedlings grew brown and faded a few days after being put out into the ground; people told us that they would die, but we persevered and hoped and moved the sickly creatures to clean earth, and at last young shoots appeared and developed into sturdy new plants. Our affection for these valiant Canterbury bells is out of all proportion to their beauty. And so we loiter in our first walk, pausing at each bush and plant, feeling down the trunk of the chestnut tree, delighting in the smell of the budding almonds.

But a veil hangs between me and my garden. I realise what it is, only when I have pulled up a few weeds and put my hand deep into the earth. For weeks past I have been moving about, leading an untrue life, with the attitude of a spectator. Now I need to do things in the garden myself, to dig, to plant. Only in this way shall I grow really intimate with it and understand it. I start to dig, but to preserve my hands I keep my gloves on. It is not long before I see that this is no good at all. The gloves themselves act as a barrier. Throwing them off, and with them the restraints and respectabilities of my recent

existence, I am at last one with my garden, and I am happy.

And what days of digging and weeding follow. Sometimes I find myself amused at my delight in it, remembering the fierce resolution of my childhood. My father mistrusted gardeners – they dig up all one's pet plants, he avowed – and would not have one anywhere about the place, so always I was commandeered to do the weeding and clearing that bored him. "When I grow up I'll never, never, never have a garden," I resolved, as day after day I uprooted daisies from the tennis court or tidied the edges of the paths. And I meant it. But now that there is no force to command me but the needs of the garden itself, I am happy with it.

There are ominous grey gaps in the perennial bed. We hope that they correspond only to hidden roots whose growth has been delayed this year by the late spring, and that soon the earth will swell above them and burst open with the push of the green shoots; but we fear that there have been heavy casualties from last summer's drought. Proudly we see that some of our plants have grown so big that the roots should be divided; this, surely, means that our garden is growing up. But even then gaps will remain, and desperately we procure new plants to fill these hungry spaces. We know that lupins refuse to grow in our grey chalky soil, but we cannot resist the temptation of getting some. The only thing, then, to do is to provide the poor plants with a bed of alien earth. Noel volunteers to bicycle into the plain with a large sack and bring back some good Oxfordshire "dirt", as the villagers call it, free of any hint of chalk. He returns with the sack full of wonderful dark brown soil, strapped across the back wheel of his bicycle like the saddle bags of Mediterranean donkeys. As he tumbles it out into the lupins' prepared bed, we feel that it almost makes us want to be lupins ourselves, so that we could be planted in it. For there is something luxurious in dark brown earth to those who live in a world of pale grey chalk. An American friend who once visited us, asked us what strange white fertiliser we had spread over our garden, and would not believe us when we told him it was merely our earth.

Now the lupins have been tenderly packed into round holes lined with manure and "Oxfordshire dirt", and we have watered all the new plants, and divided the overgrown roots of veronicas and Michaelmas daisies, we are feeling less reproached by our garden. We can, in fact, pause to enjoy it.

In the vegetable plots straight lines of green appear. The wire guards can be removed from the first peas that are showing; broad beans have pushed their heads above the ground. Darville, who comes up from the village pub to garden for us, plants potatoes; for it is Easter time and potatoes should always be planted on Good Friday or as near it as possible, be Easter early or be it late.

And then rain falls, a gentle "growing" rain, as the villagers call it. They look upon it as a friend. To love rain one must live in the country. It falls for several days and the plants strengthen and swell. The faces of the villagers glow as we meet them and discuss it. I listen to it as I lie awake one night. There is at first silence, for the rain has ceased. This silence is so dense as to seem to be black. Then comes the silky rustle of soft rain like the sound of a gentle wind in a ripening cornfield. Some while after comes the steady drip of the rain from the pipes into the rain-water tank. From time to time I hear a far off owl, and once or twice some bird seems to turn in its nest and sleepily grumble. From up the hill comes the raucous sound of a turkey. Soon dawn shows silver grey, and as at a signal several larks rise into the sky, singing in the rain. The grumbles of the sleepy birds cease, and suddenly with the lark the gardenful of birds bursts into song. This birds' song at dawn is so punctual and so simultaneous that I wonder at its ordered pattern. It as suddenly ceases, with no straggling song of any individualist bird breaking the clear-cut silence. I get out of bed and go to the window to search for this well-trained choir, but no one is visible. I am rewarded, though, for I see the orchard gleaming white with budding daffodils, and I know that it is wrong and wasteful of us to look at things only in their habitual light. The half-opened daffodils in the grey dawn have a tender beauty that they lack in daylight.

The first swallows arrive and twitter. We grow excited as we see them prospecting in front of the house, for we have never yet been honoured with a swallow's nest. We feel that our blue plastered eaves should attract them, and remind them, on dull grey days, of Mediterranean skies. Last year they got so far as to teach their young to fly from our eaves and our telephone wires; but no nest appeared. What made us feel particularly sore was that they built near us in a house that we considered architecturally inferior to our own. There seems no way of informing a bird that she will be welcome and cared for and left undisturbed. But the prospecting today has a look of serious intention.

There is a strange order in things this year; for here is the first swallow with us while half the daffodils are as yet only greenish yellow buds. As the seasons are early or late, so do we get varying

combinations of blossoming. Last year, with its especially early spring, brought the flowering American currant into bloom with the plum, and scattered the almond blossom upon shaking daffodils. Now, today, daffodils are coming out as the apple blossom still shows pink. The trees follow the calendar more steadily, except that the young trees seem always to come into leaf much later than the full grown ones.

As the month draws on, there is a sudden rush of warmth and in a

day or two the garden is a changed world, as in a fairy story where a spell has been suddenly lifted. Barely budding trees thicken with green, the spinach is rampant, and the rhubarb, that a few days back was pale and stubbly, is like an enormous tropical plant. The cherry trees that should have been in bloom for Easter, now burst into festoons of hanging blossoms. Yesterday the buds were still tight and colourless. Pear trees are heavy with flowering clumps. Sticky buds of the chestnut change and stretch out woolly and open with a few hours of heat and sun, showing their blossom as a hard point of pale green. Bees bumble, spiders run in the grass and ladybirds glow everywhere. As I stand, for luck, on the grass I hear the first cuckoo. After having been held back for over a month, all the

things tumble over themselves and each other in their urgent rush
to bloom. Has there ever before been such a rush? Frantically one
tries to take it all in, but one misses many subtleties in this glut of
blossoming. The primroses and daffodils that should have had place
of honour a few weeks back now excite us somewhat less, because
the lilac buds fatten so visibly, and anemones and violas flower. This
telescoped spring wastes much beauty.

Darville comes up one morning as though he has something
specially important to say. I have my suspicions, and expect it is
about the cucumbers; and I am right. It is his pet subject. Last year
we bought a cold frame and had exciting plans for it. We would
shelter and force many delicate seedlings and grow rare flowers. But
one day we came into the garden and found that Darville had filled
it with cucumber plants. Lushly they swelled and grew, occupying
the frame for the whole summer, needing to be covered from the
midday sun and watered at evenfall. Actually we did have a good
many cucumbers from it. But we swore we would never let it happen
again. Those cucumbers had cost us much in time and thought and
worry, and wasted our precious frame. Elaborately this year we
have filled the frame with boxes of seedlings and imagined we were
safe. This morning Darville tells me that the seedlings should now
be put in the open to harden off, and that this will leave the place
free and ready for the cucumber plants, which await us down at
his house. So we are lost once more, for we haven't the heart at
this point to go against him. The only way of escape would seem
to be to go to the expense of a second frame. But why is there this
amusing snobbery about a cucumber frame? We are gentlefolk in
Darville's eyes by reason of our cucumber frame; and as such are fit
employers of labour. It was not yesterday that it became a sign of
social respectability, for Trollope had already noticed it. The tomato
runs a close second to the cucumber as a token of gentility. Perhaps

the second frame might be needed for tomato plants? Who knows?

As one suddenly opens the door into the garden there is movement everywhere of big birds flying off. The place is full of excited, emotional blackbirds and thrushes. I have found a hen blackbird sitting on her nest. She is in a tangle of boughs of a crab-apple tree in the hedge and so intent is she that I can go close to her. She sits so immovable that she might be carved in wood. Even her eyes remain motionless. There is some unanalysable emotion in her pose, her head thrown well back, her beak pointed upwards in a challenge of patience. Her breast is puffed forward and full, her body fits the nest all round, that her eggs may be kept warm. At the same upright angle as her beak sticks her tail, outwards from the nest. She seems so proud and tender and gentle; and still her eyes are motionless. I wonder if she is petrified at my nearness, or whether the instinct to keep the eggs warm is stronger even than the instinct of self preservation. I am deeply moved, although I cannot explain to myself why it is so. I feel as though I have been privileged in seeing her.

A few bushes away is a hedge sparrow's nest. As I go near, she flies off, disclosing one blue egg and two nestlings that have been hatched this morning. On the tips of the hedges birds sing. The hedges themselves shelter warm nests.

MAY

"AND WE'D SOAK OUR GARLANDS OVERNIGHT, and they'd be fresh and ready ever so early next morning. And when we'd got to the houses we'd sing:

Good morning, ladies and gentlemen,
We wish you a happy day.
We've come to show you our garlands
Because it's the first of May."

Annie is wiping dishes in the kitchen. Outside in the garden everything glows. It is May Day. But Annie is reflective, and her reminiscences trickle on.

"There's not much call to be here in the country on May Day nowadays. Nothing happens. Why, the schools don't even give the children a holiday as they used to do. When we was children, we used always to go out with our garlands to all the houses. We'd pick every kind of wild flower and blossom and even ask people for flowers out of their gardens. Me and Maggie Oliver and Jenny and Martha Lacey would wear our garlands on our heads and round our necks, and we'd carry long poles, twined with flowers; and we'd have a doll in a chair on two long poles, and we'd dress the doll up in white and cover it with blossoms and garlands."

Annie nods towards the window from which we can see a royal clump of crown imperials flowering on their tall straight stalks.

"The Queen would have one of them at the top of her garland. We always called them Crown of Pearls. Haven't you noticed, when you look underneath them, they seem to have pearls inside? We were happy doing all that. I wonder why they never do it these days?"

One does wonder.

There are days in the round of the year that hold everything in the cup of their hand. We have one or two of these days now, in early May, with sunshine and swallows, apple blossom and cowslips, lady's smock, the cuckoo's call and the first buttercups. All day we work to birds' songs; blackbird and thrush, chaffinch and linnet, they are never silent for long. The skirts of the hedges froth white with cow parsley. The blossom is so thick on the trees that there is no dark of bough or branch; seen from a little way off it looks like linen flung over great bushes to dry. White predominates now. Earlier in the year flowers were chiefly yellow, and later the gardens will glow with pinks and reds. I wonder whether there is any scientific reason for this control of colour at certain seasons of the year? Is it planned for the sensitivity to colour of the especial insects that are to pollinate the flowers?

The orchard shines golden. Daffodils withdraw and give place to dandelions. The dandelions teem under the encouraging warmth of the sun. Beautiful as a blaze of colour, and beautiful in the form of the individual flower, they should be allowed by reason of their beauty to remain. Experience, however, has hardened us against them. We slaughter them with hook and shears before they shall seed and cover the garden. We place poison at the heart of each plant; but in spite of all this their numbers increase each day. For we have against us uncountable roots of many years' seeding, when dandelions flourished unchecked in the rough meadowland. They are strong opponents. The massacre of dandelions is a peculiarly satisfying occupation, a harmless and comforting outlet for the destructive element in our natures. It should be available as a safety valve for everybody. Last May, when the dandelions were at their height, we were visited by a friend whose father had just died; she was discordant and hurt, and life to her was unrhythmic. With visible release she dashed into the orchard to slash at the dandelions; as she destroyed them her discords were resolved. After two days of weed slaughtering her face was calm. The garden had healed her.

Suddenly, after this stretch of exuberant heat had brought the garden to the flowering of early summer, a frost struck it. Of all the cruelties of the year's weather, this May frost is the worst. The recent warmth has so encouraged the plants that they are especially vulnerable in their tender growth. As we look at the frost-shrivelled leaves of the hedgerows, we think of nests full of eggs and fledglings. Many a mother bird these nights must have spread her wings wide and puffed herself out to keep her youngsters warm. Stinging hail follows the frosts, beating down seedlings and young shoots of plants, smiting the backs of sitting birds through the shelter of the bushes. Darville is full of it, as he comes up to attend to his cucumbers. He assures us that he had expected it. "Why, they say about here that as many fogs as we have in March so do we have frosts in May. And it seems to be true. Old Mr Saxton was talking to me the other day about this lovely warm weather. 'You wait,' I says to him, 'you wait until after the 17th May. Why, last year I remember the 17th May my taters were all cut

off.' And sure enough there was ice on the water butts last night, and it looks fair to continue, too." It has browned the apple blossom and blackened the hearts of the strawberry flowers. But even so it is a mild frost compared with one we had late in May a few years back, when the earth was white early in the morning as with a snow storm, and the young leaves were frozen black to the trees. For the rest of that summer the wreckage of the frost was evident all about us and

ash trees stood leafless, a continual reproach.

One morning as we hoe between the rows of young carrots, someone calls to see us. He is from the local Horticultural Society and asks if we will become members. I feel absurdly proud, more so than if I had been offered some distinguished world-honour. There is a freemasonry of the earth, and I like to imagine that some special qualification is needed other than the mere possession of a garden in this neighbourhood, and that there are initiation rites more exacting and requiring greater prowess than the annual payment of a two shilling subscription. As I read the schedule of the classes for the Annual Flower Show, I wonder if we shall be worthy of entering for any of them. Would our huge sticks of rhubarb take a prize? If the poppies in the back bed, that we have so liberally fed with bone meal, continue to grow like giants, might we not send some of them, length of stalk and all? First prize, three shillings; second prize, two shillings; third prize, one shilling: but how proud we should be! We feel tempted to try our luck, though we know it would be hopeless. There are against us all the cottagers in the plain with their rich earth and sheltered gardens. I look more closely at the classes: vegetables, carrots, onions, beets, cut flowers. I pin my hopes on the class for the finest buttonhole for a gentleman; perhaps I might manage that, or the best arranged bunch of cut flowers?

My mind goes back at this point to a flower show of my childhood. My brothers had made great fun of me for my choice. Liking their shape, I had submitted a bunch of pink mallows. My elder brother, always of an extravagant turn of mind, had assembled the best roses from the garden and had even pilfered some of my father's treasured lilies from the greenhouse. Beside his gorgeous bouquet, my common little mallows looked apologetic. But I was defiant and persisted that my flowers had beautiful shapes; at seven years of age one is not yet a social snob – not even with flowers. It is extraordinary how many people do have a snobbish attitude towards flowers. Quite a sensitive friend glanced at a jar of cow parsley in our open grate and said in surprise that it looked really pretty, adding in a voice of smashing finality: *What a pity it was a weed*. Annie, however, who looks after us, is a good corrective to all this. She has no feeling of class towards flowers. One of the nicest things I can remember is the bouquet she made for us one summer. We were giving a party in London, and pillaged our garden for flowers to take up with us: roses and delphiniums, lilies and anchusas in rich abundance. Annie arrived that morning with an enormous bunch of wild flowers that she had gathered on the railway line; and towering among the campions and moon daisies were rust-red spires of dock.

"They'll be ever so nice up there for the party," she said. "They'll make all your friends wish they were in the country, too."

And the bunch of wild flowers honoured the party.

It would be difficult at this time of year not to see beauty in weeds. The ditches in our lane are milky with the white nettle, that exquisite flower with the form of an unsophisticated orchid. In the orchard the sequence of flowering weeds runs its course. As dandelion fades, it gives place to brown plantain with its lace ruff, speedwell, purple vetch, buttercup and cow parsley. Spikes of the beautiful wild salsify stand among the growing grasses, closing their flowers at midday: goat's beard, as some call it, or John-go-to-bed-at-noon, as it is known to others. Along the hedge-bottoms gleam the cuckoo-pints, challenging the darkness of the hedge shade with pale green staves, flapping their limp petals over themselves as their time of flowering draws to its close. Was any flower the possessor of so many names? Cuckoo-pint, priest's pintle, lords and ladies, wild arum, good King Henry. There is as much charm in the names of our wild flowers and weeds as there is in the flowers themselves. Who can resist the Lincolnshire name for the wild pansy: meet-her-in-the-entry-kiss-her-in-the-buttery? Or Venus's looking-glass, or Jack-by-the-hedge? One can understand the origin of some of these intimate names, such as Queen Anne's lace for the frothing filigree of the cow parsley, or son-before-the-father for the autumn crocus, where the flower emerges from the bulb and blooms six months before the leaves are thrown up. But I wonder very much who the good King Henry was? As one comes upon these names, one feels even more strongly the pity of it that the old names for our flowers are fast being buried

under the Latin titles. Hardly a soul now talks of his snapdragons, yet how much more beautiful that name is than antirrhinum? Pinks have become dianthus, and in cottage gardens the humble marigold is transformed into the calendula.

We plant out seedlings raised by ourselves: violas, zinnias, cosmos and sunflowers. With a proud gesture we give our surplus seedlings to neighbours; for do we not thereby show that our garden is developing? A year or two back we clutched at everything we could lay hold of.

And then, just as I am getting proud of my garden and imagine it to be doing great things, I go to a garden in the Thames Valley. As we drive from our hills, we notice how, on nearing the Thames, the vegetation gets steadily richer, until the gardens about us are flooded with colour. Where we had left behind colourless iris buds, here the irises bloom richly; where our Oriental poppies were in tight bud, here the gardens are splashed with scarlet. I tell myself that I do not feel moved with this pageant of bloom, and wonder why it is. Do we perhaps get so much attached to the plants in our garden that we love the particular plant rather than the type, knowing each turn of stem and bud, each rib of structure? Or is there something in us that enjoys the hard fight we have had for our plants in the cold chalk of our hills, so that we resent this easy profusion that shows no signs of battle? Or are we frankly jealous for our plants, as a mother would be if other children should have better food and education than hers?

I feel that we must watch ourselves lest this jealousy should slide into a possessive attitude towards our garden. We ought to be custodians, not owners; it should be our privilege to help the living things in our garden. A really good man should want to tend a garden, even if it is not his own; this is the decisive test.

As I look at the static richness of these flaming masses of colour, I feel that a man's character shows as much in his garden as in anything else. The aim of a garden is usually evident. I know one man who concentrates on rare plants, merely because they are rare, never minding how uninteresting they may be. Another grows rare shrubs. Yet another seeks the original types of flowers to the exclusion of fashionable variants, while a friend of his is interested only in foreign plants brought back from holidays. The most usual concentration is on the rock garden: there seems to be a special magic for people in the scattering of white pieces of rock about their earth. The care for miniature plants is said by some caustic psychologists to satisfy a thwarted maternal instinct. It is certainly true that it brings out any protective tenderness that we have in us. Large plants, we feel, can look after themselves; but these small creatures need our especial care. As for us, we must resign ourselves to growing only what will thrive on our chalky soil. It is the price we pay for the beauty of our hills.

Before I leave the Thames Valley I visit the Berkshire scenes of my childhood, where I lived with my old aunt Sarah. I feel I belong here much more than as yet I belong to our garden in the Chilterns. I wonder how long it takes one to feel entirely possessed by a particular part of the earth. For there is no doubt that one is possessed by the earth rather than oneself possessing it. I unfold myself to the richness of the Berkshire soil and the oak trees and thick lanes. For this had been my sole contact with earth in my youth, my one escape from London. From Aunt Sarah I learnt the rhythm of the garden's year. In her garden I performed those annual rituals that are necessary for happiness. I remember them as they followed upon each other through the months: I saw the first snowdrop, and plucked the first primrose. I heard the cuckoo for the first time. I listened for nightingales on May evenings. I picked a strawberry warm in the sun, and climbed ladders to gather Aunt Sarah's apples for her. The lesser rituals would be spaced between these: watching for the first Christmas roses, the annual journey to the fritillary fields to pick these evil, un-English looking flowers; the finding of birds' nests in the coppice. It is through rituals such as these that one belongs to the earth.

Back in our own garden, we marvel at the change that has come over it in two or three days. The flowers are precipitate in this heat of early summer. I pick a bud from the blazing clump of Oriental poppies and bring it into the house. Before breakfast it is still encased in its sheath. Now it has burst its bonds, and after only half an hour is red and crinkled and mature. And now again, two hours later, the fully developed poppy has already dropped a petal. Darwin tulips

stand tall and gay in the flower beds; some of them so inky black that they remind me of my unsuccessful experiment as a child, when I injected ink into a tulip bulb and waited in a frenzy of impatience the winter through for the blooming of what I hoped would be the first blue tulip; but no tulip whatever showed above the earth that spring.

"Coo! There's thunder about," says Annie. And this sentence is as much a sign that summer is here as the budding of the dog roses in the hedges or the swarming of the bees. Each morning as she arrives she will greet us so. It will be her chorus to life until the chrysanthemums break into colour in the perennial bed. It is a pity she has no chance of being a pagan; for with what open arms would she embrace the religion of Thor. But though she prophesies thunder, nothing happens. Inky clouds cover the skies, and there is tension in the air. The terror of a second year of drought enters into us. We feel we cannot again face the misery of our parched plants. But the earth is dry to cracking point. Gardeners say it has forgotten how to rain, and everyone despairs. The vegetable garden looks happy to the casual observer in its definite stripes of various greens: pale rows of lettuce, red brown of beet, feathery lines of young carrot, frills of turnip and spears of salsify. Peas flower and climb in a tangle of beech branch stakes. Strawberries are in blossom. But lettuces are tough, and carrots are tiny, the spinach is running to seed and the spring cabbages refuse to "turn in to" hearts. The water butts stand nearly empty. Each day now we watch the sky for Annie's promised thunderstorm. But day succeeds day in parching, rainless heat. The orchard grass grows high, hiding more and more the birds hopping in it, and the insects in its depth. In the ordered procession of flowering weeds it is the hawkweed's moment. The tall stems bend under the weight of the many pecking sparrows that alight on them. The golden flowers glow in the strong sunlight. The garden thickens. The paths seem smaller and narrower, as the bounding hedges swell. As Annie goes to the kitchen garden to pick vegetables, she is now invisible from the house.

Through these blazing days a crimson-stained linnet sings in a poplar tree. Annie, who has a flare for birds' nests, finds the home of the little hen. She is sitting in the same hedge that she occupied last year. It might almost be among the same branches. Hour after hour ripples the shrill, trilling song of her mate, cooling the air as though it were a stream of icy water. The dull little hen linnet sits tight and never seems to leave her nest.

We are unlucky this year with our thrushes' nests. One was abandoned before the eggs were laid, and now we have found a second one, hidden from sight in a cleft of boughs of one of the beech hedges. Neatly lined with cow dung, it contains four eggs. For days we watch the nest from a long way off, but never do we see the mother bird. At last I look more closely and there, partially covering the eggs, are dead beech leaves that have fallen into the nest as the new leaves above and around it pushed off last year's growth. Knowing now that the nest is deserted, I touch the eggs. They are cold as stone.

But my blackbird in the crab-tree is a successful mother. In her nest are three young birds, still naked; their heads and eyes and beaks are enormous; amusing tufts of hair sprout from head and back. As I take one from the nest and place it on the palm of my hand, it stretches its head from time to time and opens its beak for food. At this stage it knows no fear and has but one purpose in life, to consume all it can get. But the mother bird, watching, makes disquieted noises from the tree. I replace the grotesque object.

Everything bears abundantly this year. The newly planted maiden fruit trees are covered with setting blossom. Reluctantly I go the round of them, picking off nearly all the young fruit. It is a heartless job and seems a rude rebuke to their enthusiasm, but it is necessary, if young cherry and young plum are to grow up strong and bear well. In the hedges the white bryony throws out its tendrils, covering the hawthorn and wild rose with its rapid growth; what does it matter how much it covers, with such beauty of leaf and tendril. All the garden plants and bushes thicken: summer growth nears its climax.

JUNE

THE GARDEN IS A NURSERY of nests and young birds. The thickness of a hawthorn hedge is the home of five lusty young hedge sparrows. As I move the leaves aside, the slight noise wakes them. It must mean food. The five youngsters shoot upwards in the nest like flames, embryonic wings fluttering, scraggy necks twisted and erect; in this ceaseless craving for food they stretch their orange mouths wide, like tulips opening to the sun. As they spring up, they give a little scream in unison. In all they do, be it of movement or sound, there is unbroken unity, as though they were puppets worked by a single string. I tickle them with a leaf, but they soon discover that I am not their parent and bring no food for them, and they subside and lie snug in the nest, interwoven in a pattern of heads and legs and wings. After a short while I shake the leaves once more, and the fledglings shoot up in the same rush as before. Soon they grow suspicious of this rustling sound, and though I shake, they will not stir. But suspicion dies easy in young birds, and I return to the nest after a short walk round the garden to find that they spring up again at the slightest stir of leaf.

We find other nests. At this time of year the hedges and trees pull at us, and we have to remind ourselves continually that it is kinder to the parent birds to keep afar off. However much we may resolve to avoid the path near a certain bush, our feet will take us there. I think this urge to look for nests is not so much idle curiosity as a desire to share with the birds the tenderness of their family life. One is ashamed when a bird shows fear, and takes it as a personal reproach. I know that I have a feeling of great pride when a bird stays undisturbed on her nest as I pass close to it. A friend of mine told me that one of the happiest moments of her life was when she visited a small unfrequented island; the sea birds knew no fear and let her stroke them, as they sat on their nests on the ground and among the rocks. One of Noel's chief delights is to tell the story of his return from Norway in autumn. Early in the morning he was on deck, and saw a chattering crowd of migrating golden crested wrens that had attached themselves to the ship and were running up and down the rigging, pecking at the masts as though they were tree trunks. Noel, who always carries pine kernels about with him in his pockets, lest he should meet birds, threw some to them; but they took no interest. He fetched bread, but even that failed to attract them. So he sat down to watch them. Soon two or three of them had clambered up his deck-chair and were climbing over his body, hopping about in his hair, and pecking there for insects! He says that the feel of the little creatures fearlessly wandering over him was one of the best things he can remember.

The young birds that walk about on the grass just now are bold, fearless creatures. So much is this the case that I even have difficulty in getting out of the path of the young thrushes. I wonder at what point these confident youngsters learn fear? Does it come upon them suddenly, or do they develop a gradual lack of trust? We would much enjoy taming them, but feel that it would be unfair to them; for one could not guarantee that the whole world would always treat them kindly, while one would have stopped the development of their instincts of self-preservation. Annie, who has a "way" with her where birds are concerned, stands at the kitchen window, feeding her family. What does it matter that the house gets cobwebby? The parent birds need vast quantities of food just now for greedy babies, and the young birds themselves are brought along to her to be fed: one mother thrush, one father thrush, three young thrushes, two blackbirds with a child, a cock and a hen chaffinch, and a sprinkling of unwanted starlings and sparrows. For Annie's hospitality does not willingly extend even to the starling, with its stupidly clumsy-looking young, and sparrows she will not suffer. The thrushes and blackbirds come within a yard of our hands in their repeated visits for food for their young. Occasionally a thrush will suddenly grow timid and shrink in size as he draws near to us, hoping thereby

to escape notice. But Annie has discovered that the birds have a special liking for cheese and cooked lentils, and each morning the wall outside the kitchen door is sprinkled with them. Their greed overcomes all fear.

On the grass walks, and all about the garden, fat young birds follow their parents for worms. Our grass walks form a perfect nursery, where, sheltered by hedges, the young birds can find a plentiful supply of worms; so, at the making of the garden, we had planned. There is no insistence in the world to equal that of the perpetual squeak of a young thrush as it hops after its mother, demanding to be fed. One can almost imagine that the mother bird must get impatient and try to elude her offspring; she must else have a quite superhuman power of endurance: but then, she has already proved this to us by her patience as she sits continuously for three weeks on her nest. I have seen a cock blackbird get really angry when his enormous speckled son stole from him an especially fat worm. I watched him as he tried to dodge the persistent youngster, who would walk after him everywhere, turning round on him and scolding him; unable to get free of him, in desperation he flew to the top of the chestnut tree, too high for the fledgling's wings to carry him. The young birds are larger than the parents and always fatter, for the adults spend their time flying from place to place for worms, and generally look very thin and worried. If it were not for the lack of tail feathers, one would hardly know youngster from parent. I am reminded of some of the large lambs that butt at their mothers for milk. There is the same selfish insistent look in the young thing, the same patient endurance in the mother.

In the top corner of the shed we find a wren's nest. It is deep and looks just like a dark hole. It is full of tiny embryonic wrens. Two have tumbled from the nest to the ground, and lie on their backs, their legs crossed over their large blue stomachs. As I pick them up, I am struck by the relics of their reptilian origin, with their disproportionately large eyes placed right on the tops of their heads, as in a frog. Their half-formed wings are like the flappers of a fish. At this stage in their development there is little in them of the grace and beauty of form of the wren.

The abundance of young birds seems endless. In the orchard noisy young starlings are being instructed in trial flights. A chatter, irresponsible as the talk of schoolgirls, comes from the row of poplars against the shed; three baby coal tits are trying their wings. They run up and down the poplar trunks, jumping off from one tree to another, learning to fly in little movements in the air. They are joined by a family of fledgling chaffinches, fresh from the nest; the newly planted plum trees are very popular with them as a taking-off point. As we watch them we decide that we should always manage to have a few small trees in our garden at this time of year. Young birds soon tire: they are as yet so accustomed to sitting in the nest that occasionally they flop down in the grass to rest from their exhaustion in the middle of their hopping walk.

And all this time, while the garden is turned into a nursery for young birds, the little hen linnet sits tight on her late nest, and we have discovered a new thrush's nest, filled with warm eggs. Courtship continues, too. Surely our blackbird is wooing again. What else could account for the deeper, more flutey note that has just entered into his song? He has never sung so well this year. And then we see him and understand the fervour of his new note. In the middle of the grass path through the orchard, shamelessly he courts his hen, dancing round her and upon her. Soon he tires and flies off, but she is willing and urgent, and chases him to the darkness of the hawthorn hedge. In the middle of the potato patch, in a valley between the ridged potato plants, the gay little chaffinch covers his demure, acquiescent hen, to make a second brood. Just now the garden belongs to the

birds, and it is with fear that we hear the particular danger cry the blackbirds give when a cat is about. Throughout these days we dash from the house at this appeal to chase the neighbouring cats as they stalk our tailless young birds; for the fledglings live on the ground and are fearless. Noel is worried, for he loves cats. Perhaps our loveliest neighbour is Banshee, a pure white Tom. Throughout the year we encourage his visits, and get great pleasure from his beauty. He will lie on his back, and roll and turn for our benefit while we dig; he rubs himself against our legs as though we were tree trunks: what are we to say to him when the blackbird cries out against him, and he prowls our walks to pounce upon our young birds? It is hard to know.

Darville comes up this morning to shade the young tomato plants from the sun. He is full of bustle about his bees. They are swarming with unusual vigour this year in this fine weather. One swarm of nearly eight pounds has snapped the end branch of his plum tree with its weight, and fallen to the ground. There must be a swarm somewhere near here today, for the garden is restless with bees and the kitchen hums with them.

The drought continues. The rainless heat brings grasses and flowers to blossom and seed in a flash of time. Never has the rose bed glowed with such colour, for the roses are happy in the sun; but they bud and blossom and fall in one day, and the garden grows exhausted with the precipitate flowering. The violas degenerate for lack of moisture, growing small before the fullness of their blooming is reached. The height of summer is here, with heat quivering over the fields and the noise of a blue-bottle in the room, bumping against the windows. The air is tight and strained with the drought. Annie assures me that we humans are needing rain just as much as our plants, and perhaps she is right. It would soothe our nerves and loosen this tension. Most of all, it would save us for a time from the reproach of our wilting plants. As the flowers rush past their time of blooming, I do their toilets, cutting off the dead stalks of columbines, Darwin tulips and irises. In this rush I can hardly keep pace with them. There is an air of heartless finality about doing this, as though one would be done with this year's flowering and hurry

along to the next item on the year's list.

At their appointed time the irises bloom and fade, in strict order of colouring: first the most usual purples, then the yellows, next the pale mauve and silver flowers, and finally our ghosts. These last are especially lovely. I had brought them from Aunt Sarah's garden; they are so transparent in their mauve and grey and buff that one is surprised at not being able to see through them to the hedge beyond. Apart from their intrinsic beauty, I am sentimentally attached to them. For there is no happier legacy than flowers. We have transplanted to our garden Aunt Sarah's crown imperials and Christmas roses, daffodils and irises, and I remember her better in them than in anything else. At the foot of the orchard, among our cultivated daffodils, stands a clump of tiny wild ones that a poet-friend of Noel's had bequeathed to him, a living memory that grows and spreads each spring.

The grass in the orchard turns fawn and brown and rust red. It must be cut soon, when the leaves of the daffodils have died off. We look forward to the mowing of the grass with excitement, and already in our minds divide the area of uncut grass in our garden, that each of us may have an equal share in the mowing. For scything is one of the great pleasures of our year. Darville smiles at us.

"Well, well, I never before heard of anyone *wanting* to mow a meadow," he laughs. "I expect you'll be asking me to help you before you're through with it."

But each morning now we investigate the fading daffodil leaves among the seeded grasses, and feel the ugly browned tops of the autumn crocus leaves in the uncut grass in the middle of the front lawn. It will not now be long.

The year's harvest begins. Clusters of empty stones hang from the cherry trees; the birds are up earlier than we are in the morning and always our cherry crop goes to them, a fitting reward for their song. The little cherry trees shake as the blackbirds peck at the fruit. Red and white currants are colouring up well. Annie brings us old lace curtains, that we may swathe the bushes against the coming ravages of the birds. Raspberries and loganberries are forming. But the strawberries are ripe. We go strawberry picking, skinning off the covering net, removing the wooden pegs that hold it down. The net catches in the topmost leaves of the strawberry plants as we tear it off. A blackbird perches on a pear tree near by, watching us with interest and ready to seize his moment should we be called away and leave the strawberry bed exposed. For it is full of ripe berries, warm from the sun. They are large and heavy and drop towards the ground, hiding among the tangled stalks of protecting straw. Like digging potatoes, it has the exciting uncertainty of a treasure hunt. Our hands feel for the soft warm fruit, our eyes search out the crimson drops in the dark shade of the plants. Slugs have been damaging the fruit, finding in the shelter of the leaves and straw the only cool damp in the garden in these days of drought. Birds have pecked at some of the strawberries, cunningly pressing down the net with the weight of their bodies, until they could dive through the mesh with their beaks and reach the sweet tasting berries. But bird and slug have left us plenty and day by day white flower drops its petals and forms tiny green berry, and hard green berry softens and swells in the sun to crimson strawberry.

We have our first dish of peas. The aisles of pea plants grow tall and the green walls are full of bulging pods. Annie says she will pick the peas for dinner, but we rush to the garden to do it ourselves, excited at the start of yet another harvest. Compared with the gathering of strawberries, pea-picking is intricate, but undramatic. There is no sudden glow of crimson, no soft warmth of fruit. It is a world of shapes, pea being distinguishable from leaf only by reason of its bulk and form. We pick by feeling rather than by sight. The pea plant is

a gentle green, deep and soft against the pale colour of the lettuces that shelter from the sun in the shade of the pea rows. Our baskets full of hard, rattling pods, we pull lettuces for salad. It is good to feed oneself from one's own earth.

Darville brings us young cabbages and sprouts, savoys and broccolis. He plants them out between the lines of potatoes. He sells them to us by the quarter hundred. As I count them I see thirty of each variety and question him. He seems amazed at my ignorance.

"Why, didn't you know that we always sell plants in a long hundred? That's a hundred and twenty. So of course a quarter hundred is thirty."

I feel rebuked.

The large perennial bed is a blaze of colour; delphinium, lupin, anchusa, veronica, bloom with unusual intensity. The red and yellow gaillardias, a gift from Darville, flower madly, in Noel's eyes ruining the colour scheme of the entire bed. I laugh to myself at the violence of his hatred, knowing that it dates back to his childhood, when a Salvationist nurse terrified him with stories of hell fire and the

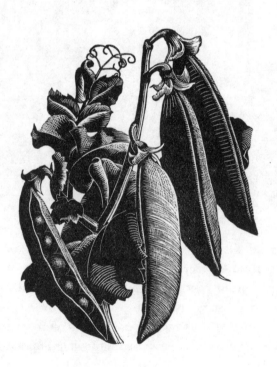

blood of the Lamb: now it is as much as one's life is worth to put yellow and red flowers together, be it in garden or house. He wants to uproot the wretched gaillardias, but I tell him that it would be too unkind to Darville. So he picks off the flowers and gives them to Annie; but the more he does so, the stronger grows the plant and the more it blooms.

We have other undesirable creatures in this big bed. Unsown poppies are sprouting up everywhere. One is lenient towards flowering weeds in the first year or two of a new garden, glad of the colour of anything that blooms. One even declares that one is striking out against snobbishness in gardens and will find room for such beautiful things as scarlet poppies. And so, for a year or two, they do bloom. But, also, they seed. And we are paying for our soft-heartedness. From now on, the wild scarlet poppy in our garden is doomed. But a far harder thing to fight is the lovely rosebay willowherb. It appeared uninvited in the very middle of the perennial bed during our first summer. We loved it, and protected it from the onslaught of Darville. We defended its beauty to people who condoled with us on the appearance of such a lusty weed in the middle of our best bed. We laughed at Darville when he said that the day would come when we should be sorry for our kindness towards it. We felt confident that we were capable of keeping the lovely plant within bounds. But we have learnt better. Its sentence of death did not come for a year or two after that of the scarlet poppy, but today it, too, is doomed. For we discovered, too late, that it is as lusty of growth below the earth as it is above, and year by year would disclose strong new plants yards away from the parent plant. We thought it might be possible to transplant some of the shoots to the wild part of our hedges, that we might still see its beauty and touch the soft silky wool of its seeds; but Darville pleaded with us for its entire destruction, and we are listening to him. It would seem that

we must bow to convention after all, and do as others do.

As we pass Midsummer Day, the hedges are littered with empty nests. Five young linnets have flown, leaving behind a shallow, untidy mass of wool-lined twigs. The mother thrush has left nothing to chance this time. Her eggs hatched, she watched her fledglings until they were ready to fly, sitting on them with their beaks protruding round her body like the barrels of guns on a battleship, or perching by them on the edge of the nest, like a nurse by a child's bedside. Annie, too, kept intermittent guard with her, reporting to us how events moved. "They'll be flown this ever-day," she pronounced at last. And that evening the thrush's nest was empty.

The loveliest creature of the garden at this moment is the mullein caterpillar. Punctually, with that accuracy in timing that is so wonderful, the cocoon of the moth has raised itself above the earth, knowing that mullein leaves are thickening. The moth has emerged, to lay her eggs on these downy leaves. And now, with the mullein plant grown tall and thick, the caterpillars are out, feasting on the fully developed foliage. They dot the plants, themselves as beautiful as any blossom, with their black and yellow markings on their delicate pale blue bodies. Anxious to watch these creatures grow, I make a pet of one, and bring it into the house on a mullein leaf. Starting as a half inch streak of speckled wool, it grows at an incredible rate. In two days of incessant eating it has doubled its size. On the mullein

plants the leaves are fretted with big holes, and the excrements of
the caterpillars cover the foliage like brown hail. Touch a caterpillar
and it will wriggle convulsively, falling to the shelter of the lower leaf
joints: for it does not intend to be separated from its food supply.
Their beauty increases with their size. My little fellow fascinates me,
and has grown nearly two inches long. Now, today, he is restless
and, in spite of a freshly picked mullein leaf, he will not eat, and is
trying to hide in the darkness of the under side of the leaf. I watch
him, as he spins a covering for himself, for he is midway between the
burrowing and the spinning varieties of caterpillar. In a few days he
has drawn the leaf tight round him and is invisible. The mulleins in
the garden bloom erect and yellow. They have served their purpose
for this year.

The heat continues, and the drought grows increasingly serious.
There is no sign of a break in this weather. Night brings no relief,
and the heat filters through the hours of darkness. As we sit with
doors and windows open, the rooms are bumping and flustered
with moths: white miller moths and brown furry ones, like squirrels,
brush our faces as we lie in bed. During the day-time the rooms
boom with bees. The world hums with summer.

JULY

THE MOMENT FOR MOWING HAS COME. We go into the shed and look at our scythe. From the fields around comes the drone of the reaper, sending us almost to sleep with the bee-like hum, now nearer, now further, of its graduated waves of sound. Even the voice of the mower as he calls to his horses has a humming quality. So deeply are we under the spell of this drone that we have hardly the energy to think of our own mowing. The scythe seems heavy as we lift it from its hooks on the shed wall. If that soothing hum would only stop, we could brace ourselves to start. We linger and pause, and look about us, eager for any excuse to delay work. Perhaps the daffodil leaves are not as dead as we had imagined last evening in the half light? But when we go to the orchard to look, we see that there is no excuse for us, they lie crinkled and brown among the roots of the grasses.

The grasses undulate in the breeze, with the motion of a slight swell at sea. As we walk round the orchard, now facing, now backing the sun, they change colour; they are pale silver fawn with the sun full on them, and darker and redder against the sun's light. And as the

men and women in a vast crowd have their unremarked, individual
beauty and character, moulded each in his own fashion, differing
each from the other in shape and colouring, so are the grasses in
the orchard composed of multitudinous varying forms, some frail
and fine, some erect and sturdy, each with its own pattern of life.
I look closer. Where the rapid glance perceives a mere shimmering
stretch of fawn, I now see the cock's foot grass, with violet tinted
flecks of pollen still sticking to its rough spikes; rye grass and vernal
grass are light against brown plantain; the flowered green timothy
grass towers smooth and erect, austere of form among the shaking,
quivering totter grasses. Shorter, in this vast crowd, clump the
"backbone" grasses, familiar to us from our childhood's game of
"Tinker, Tailor." Pale meadow soft grass and meadow poa add to
the waving buff, a background to some still blossoming pink vetch.
Aristocrat amidst this multitude stands the lovely yellow melilot,
like a beauty in the market place. Burnt spire tells of seeding dock
that we have overlooked in our weedings. Bladder of white campion
looks strangely smooth against fringed grasses. At the far end of
the grassland grows a white clot of moondaisies. We did not dig
the orchard land when we made our garden, merely cultivating the
ground that surrounded each fruit tree; so today it blossoms with
stray lucerne and white clover, heritage from the days when it was
pasture.

Beneath these towering, tapering grasses, in the thick tangle of
undergrowth, creep little yellow hop trefoil, and scarlet pimpernel.
The gilt downy seeds of the goat's beard lie low in the grass, seeking
the earth. Below this carpet of flowering weeds are the dark homes
of insects. Moss covers the nest of the wild bee, which hums in the
short stemmed clover. Butterflies are a brown and blue mist among
the grasses. Ripened vetch pods burst in the heat with sudden
crackles.

Whether it is the unceasing drone of the reaper, soothing me till all action seems impossible, or whether it is genuine concern for this world of grasses, still I hesitate to mow, telling myself that it is wrong to destroy such beauty.

And then Noel sharpens the scythe. The rasping sound of the whetstone on the blade wakes me from my languor, stinging me like a plunge into icy water. We must start mowing.

But which of us is to begin? Each hands the honour to the other, knowing what excitement lies for us in that first crescent of scythed grass. We toss, and Noel wins.

He starts on the front orchard, a few months back dim with a carpet of bluebells. With the first cut into the grasses a host of moths and insects flutters out from the shelter of the roots, and a mist of pollen rises into the air. He mows, bent over the scythe, swinging his body from right to left with each new swathe. As he cuts, it would seem to be one of the easiest things in the world, even as it is one of the most beautiful. For there is the same beauty in the shape of the scythe that is in all fundamental things, where shape has been determined by need; so does one think of the lines of a boat, or the curve of a waggon. The sun shines on his back and arms as he swings the scythe, a figure of clear-cut light and shadow. The grasses are so dry this rainless summer that it is like cutting wire; even the dew of early morning soon vanishes and the resisting grasses blunt the blade. Noel has to stop often to sharpen the scythe. He holds it upright, stroking the blade with the whetstone along its length on either side. He stands thus ennobled, for there is no pose that is not lyrical and rhythmic when it is tied to the sweeping lines of a scythe.

Crescent follows crescent, and the front orchard takes on a new pattern as I watch. The blade cuts low at the stems, the ground is a tangled mass of prostrate grasses, their identity lost. Timothy grass and lucerne, cock's foot and the lovely melilot have been levelled by death. Through the heat there comes to me the regular swish of the scythe, silky as the rustle of a woman's skirt; the stroking sound contrasts with the sharp rasp of the whetstone on the blade.

And then it is my turn, and I learn that it is not so easy. The movement that looked simple is the balanced play of all the muscles of one's body; hip and shoulder, backbone and wrist must work in harmony that I may stroke the grasses to their death. Muscles in my waist assert themselves, resisting the regular wrench that I give. In the hot sun sweat drips from my face and body. The orchard looks boundless. At first I hold the blade too high and it fails to catch at the grasses; next I lower it too much and it rips up the ground. The point of the blade seems attracted towards the young tree trunks and would spike them if I were not careful. But soon I recollect the exact movements, and my body obeys me. My scythe moves on a perfectly level plane; I cut the same segment of a circle with each swing, so that as I carry the grass with the blade it falls from it in a perfectly straight line. By my side on the left grows the ridge of mown grass. As I walk forward in the orchard, I become the actual shapes I am making. I forget the individual grasses I am cutting; the beauty of the yellow melilot is destroyed unnoticed. All consciousness has left me; hypnotised, I am happy in making shapes on the ground.

By noon the front orchard lies cut and flat, and a sweet smell comes to us as the sun turns the grasses to hay.

Blazing day succeeds blazing day, and our garden changes its form and colour. The grasses in the front orchard were the first to fall; they are followed by the back orchard and then, in turn, by the grasses at the lower end of the garden. And each day we become more one with our scythe, understanding the curve of its wooden shaft and the exact position of its handles. Our waists and backs ache less, our arms and legs grow burnt and yet more burnt. The more we mow the less do we concern ourselves with what we mow. Mechanically we move across the orchards, swinging the scythe before us.

At the last swathe, we turn and look at our work. Before us lie straight ridges of cut grass, colourless in the distance where we had first mown, greener close at hand where the grasses are not yet dead.

Darville comes up and smiles at us, knowing well that it was pride that made us finish our mowing unaided. But he is wise enough to say nothing. Tactfully he tells us he has arranged with the local farmer to cut the grass in the piece of meadow we have lately bought across the lane. To-morrow the drone of the reaper will be nearer than before; but we do not mind now if it should lull us to sleep, for

we have mown our orchards.

There follows the play of wooden rakes and pitchforks, and we rake the ridges of hay into mounds. Darville staggers across the garden with a pile of hay large enough to hide his shoulders and trunk. He is carrying it to the bonfire, to be burnt. For we have waited for Tom, the cowman from the farm down the lane, and he does not come.

"He'll never bother to fetch it," says Darville. "We all know him. He's all come-day, go-day. I'm going to burn it." So the bonfire smokes and flames for days, and my feeling of distress at the waste of the hay is outbalanced by the sparkle in Noel's eyes as he thinks of the strength that will come to his flower beds from the salts in the bonfire ashes.

The excitement of haymaking takes our minds for the moment from the pathos of the flower garden. For the drought continues, and still the skies are clear. The heat is terrific and strikes like a knife. "The weather gets hold of you," says Annie each morning as she arrives. "It's what everybody you meet says – the weather gets hold of them. And it's thundery, too." Even at tea-time we have to rush into the shade, stabbed by the sun. The flowers and trees are defenceless, and we dare not ourselves tend them during the heat of the day. Water butts are dry; there is a shortage of water everywhere. As we pour a bucket of bath water on to the roots of a wilting tree, the dry earth sucks it up, like a hungry animal at a food trough. The leaves of the sunflowers droop, hollyhocks flag. Over everything creeps and twines the little pink and white bindweed, revelling in this continuous sun and unmindful of the drought. The smell of warm strawberries rises from the netted bed as we pass. But currants are pippy and do not swell, so that even the thirsty birds disdain them.

This same heat forces the flowers. Only two or three blossoms remain on the foxgloves, at the top of tall stalks of swelling seed-pods. The lilies and roses rush through their blooming, their lives short and magnificent. Flowers are so early that one wonders what will happen in the space of time between now and the end of the year; for hollyhocks are already out and there are buds on the early chrysanthemums.

It is good just now to walk round the garden in the late dusk, when the sun has set and the grass is cool. The half light is kind to us, hiding from our eyes the reproach of wilting leaves and cracked earth; else our walk could not be peaceful. The darkening garden is soothing. Moths brush against our faces as we loiter. Owls awaken and the first bats flutter, dark against the  pale green sky. Banshee, our neighbour's white cat, slinks along the end hedge, emerging out of the darkening grasses; as he passes, a frightened blackbird cries in alarm. In the blackness of the flower beds everything is formless, but lilies and Canterbury bells gleam white, with that especial beauty of white flowers at night. The scent of the lilies comes to us from a long way off, blending with the smell of the night-scented stock under the windows. Earwigs drop upon us from little trees. The hedges rustle with disturbed birds as we pass. And on our walk we meet Cornelius, our hedgehog. Dark against the dusk, he is hard to see, but my foot knocks against him, or sometimes we notice his shape breaking the white circle of milk that we put out for him at sunset before he starts work. In spite of his formidable snorts and his forbidding frowns, he is a friendly creature.

In the midst of this gentle walk round the garden, we are drawn up with a jolt. Noel's foot touches something on the front path. We lift it. It is one of our young thrushes, dead, with its eyes pecked out, a victim of the little owl. Hurriedly we bury it, hoping that the mother thrush has not yet seen it, and will never know. In vain did she keep unceasing watch over her young. The ugliness of this discovery jars on us. It seems so impossible to understand the emotional life of a bird. How, in the extreme terrors of its existence, can it sing so happily? There is no hint of fear in the song of any bird, nor does terror mar the beauty of its life. It seems able to live entirely in the moment, isolating its joys and fears. I think of the affection of two coal tits, feeding each other with nuts on the hopper. Into their lives there entered none of the ugliness of life. Noel tells of a hedge sparrow that he saw one day, singing happily on the top of a bush as a hawk fell, like a plummet from the sky, and would have pounced upon it if Noel had not clapped his hands and frightened the small bird. He says he remembers the terror in the movements of the sparrow when it saw the hawk. Yet that hedge sparrow started singing again, a short while afterwards, as happily as before. One is forced to face this fact of ugliness in one's garden. Outside the gate, in the newly tarred lane, we found a fledgling imprisoned in the hot tar, its wings clogged. In the raspberry bed I watched an army of ants attack the emerging head of a worm. Even the thrush or the blackbird can look cruel as it tugs a worm out from the hard earth.

But is one sinless oneself? The greatest worry of the garden to me this year is the ant heaps. Never have there been so many. The cold frame was overrun with ants in their thousands, explaining to us the disappearance of all our seeds. The pansies and violas are eaten by them; they swarm on the grass wherever one may happen to lie. In one flower bed alone I found twelve ant heaps, each one a wriggling brown mass. Hardening myself, I have destroyed them with boiling

water. But all that day I imagined that ants were swarming over me, and as I went to sleep I dreamt of them. Tossing in my bed, I could not escape them; they covered me, and questioned me: Who was I to say that the garden was made for me, and not for them? How was I to know that in the scheme of things I was more important than they were? How, in fact, was I so certain that I was a more highly developed creature than the ant? They ate green flies off the fruit trees; what did I do of equal value? They kept the earth friable; what did I do comparable? Their ordered society seemed superior to our own; from what they could see, we were astonishingly indolent. The world, they argued, had been created for them, and for them alone. "Yes, but," I interrupted, "if I leave you, shouldn't I leave all the wireworms and the leather jackets and the mullein caterpillars and the slugs?" The ants grew angry. That one should compare them with leather jackets and wireworms! They covered me yet more thickly, and I remembered the live worm in the raspberry bed, and cried aloud and awakened. But I try not to discover any ant heaps now, as I hoe the flower beds.

Then, as we are nearing the point of exhaustion with the drought, it

breaks. The air weighs heavy on our senses, and clouds appear behind
Bledlow Ridge. During the evening we hear the first raindrops. It is
a blessed sound, and though the drops are absorbed into the caked
earth even as they fall to the ground, yet with them comes relief. The
night brings thunderstorms, hurling and crackling above us in bed.
Rain pelts down. Quivering with excitement, I lie awake and listen;
and then comes the sound that I have not heard for months past: the
gurgle of rain-water down the pipes into the water butts. Released, I
fall asleep to the noise of the slashing of rain on the windows.

It is next morning that the real excitement starts. We look out of
the window and see the water butts brim full. The earth that has been
white and hard for weeks past is dark grey, like the colour of heavy
smoke. The whole garden is soaked and the grasses are cold and wet
to my feet. The sky is grey and low, heavy with more rain; and the
storms roll after one another, darkening the earth still further. The
lawns are covered with hopping birds, eagerly feeding on the rain-
loosened worms; for no worms have been seen for weeks past in the
iron-hard soil. No longer is there any need for us to pour water into
the wheelbarrow, that the thrushes may use it as their daily bath.
No longer must we feed birds that could find no worms. Their food
supply is once more assured. Slugs have suddenly appeared, and in
the shade of the lavender bushes I find one large snail. It is the first
one I have seen this summer, and I wonder where it has been hiding
all these weeks. There is a rustle in the grasses at my feet. A frog
hops across my path; it is the first frog, too, that I have seen since

the drought. It troubles me to imagine how it has been living in this rainless year.

A blackbird listens for worms. As I watch him and marvel at his sensitive hearing, I think what a raucous, strident thing we humans must have made of life for the birds. Ears that can hear the wriggle of a worm must be hurt by the roar of a car or an aeroplane.

But we have a short time of release from the drought. The thirsty earth has sucked the rain deep into itself, and after a few days the garden is parched once more. The hoe beats at the earth as though it were grey stone. There is to be no enduring respite this year. And the garden grows tired. Colour leaves it. Most of the soft fruit bushes have given up their harvests: raspberry, strawberry and currant, all now are mere clumps of green leaves. Peas and roses are exhausted. Marrow flowers fade and bulge into marrows. The year is growing old very fast, the garden is bored. Only the lavender bushes bloom with exuberance, alive with bees by day and moths by night. As the year treads on, different weeds appear: milkweed and thistle, wild carrot and shepherd's purse have supplanted groundsel and plantain. The shorn orchard lies brown and parched: no rains have endured long enough to bring life to the starved after-grass.

And into this fatigued garden creep the first signs of autumn. For plums have started turning colour, and today's wind has tossed to the ground one rosy, mature apple.

AUGUST

"It was nothing but faint-heartedness that stopped you and me," said Darville. "There were no apples anywhere near as good. Why, that little branch of them, over there, would have got the first prize, and no mistake."

Darville is tying up the tomato plants, and as he potters, he meanders on about the local Flower Show. The drought had destroyed him this year and he had sent nothing; his onions that should have been a foot round were only ten inches, and he feared the shame of showing them.

Our little tree of early eating apples is certainly beautiful, bridging summer and autumn with its branches of crimson fruit. For there is no doubt that autumn is upon us, bracing us with its sharp, stinging air and relieving us of the languor of this blazing summer. The young mountain-ash tree in the front orchard is weighted down with its burden of red berries. They reach their height of ripeness for one day, and early next morning I look out upon the little tree and see it a vibrating mass of starlings, eating the berries; in an hour the tree is stripped of colour and the small branches stand upright, relieved of their burden. Sure sign, too, of autumn is the appearance of spiders' webs; festooned with dewdrops, they join raspberry cane to raspberry cane and plum to plum. We find spiders everywhere, heralds of the coming days of moisture. "There goes one as big as a week," says Annie as she dusts the sitting-room.

And the moisture does come. This time it would seem that the drought has broken for the year. The grass grows green once more, and flowers gain courage for a second blooming. Beneath the degenerate, small violas that I raze to the ground, I see a host of tiny green shoots, prepared for further flowering before the year grows cold.

But it is the beauty of seed pods that strikes me just now. The rushed blooming of flowers is over, and colour of flower has given way to form of seed pod. It is so wrong to think of the beauty of flowers only when they are at their height of blooming; bud and half developed flower, fading blossom and seed pod are as lovely, and often more interesting.

Of all seed pods, those of the carnation poppy are the most beautiful. From the moment when the petals of the flower unclose and expose the pale centre of the poppy head to the last stage in its life when the seed head lifts its cover and, like a pepper-pot, scatters its seeds on the ground beneath it, it is full of subtle changes of form. The pure geometrical shape of its cart-wheel lid, with regular spokes radiating from its centre, is evidence of the severe forms that abound in plants. If we could more often see plant forms under a strong magnifying glass we should be amazed at the regularity of their shapes. There before us would be the visible basis of much pattern, the origin of historic design. As they drop their petals, the poppy heads swell each day, some blue grey, some with a pale bloom on them such as one finds on a damson, some a bold bronze colour; and as they swell, the shape of the inner seed partitions accents the bulging bulk of the pod in a regular rise and dip round the form. As they grow yet older, and their seeds have ripened, the tips of the wheel-spokes lift and the lid raises itself up, exposing a circle of minute archways. Out of these archways drop the ripened poppy seeds, for next year's flowering; into these archways step the earwigs. For split open any ripened poppy head and the chances are that you will find a perfect earwig

nursery. It is dark and it is cool; it has an abundant supply of food in the very bedrooms themselves. Another creature who is fond of the poppy head is the little blue tit. The poppy bed seems suddenly to have blossomed again in blue and yellow as he visits it. The poppy heads shake as he alights on them, spilling on the ground the very seeds that he seeks.

Harvests rush upon us. No sooner are the black-currants finished, than the gooseberries ripen in golden globes. Regardless of the pricks and stings of plant and insect, we sit on the grass by the side of the gooseberry bushes and pick and eat. It is the one time in the whole year that we know that we are greedy, and revel in it. The vegetable marrows are ready for picking. We love the beauty of the plants. With their exotic orange flowers and their twisting leaves and tendrils they fling themselves over the rubbish heap, gracing the home of Cornelius, our hedgehog.

On a perfect August day I lie in the grass under our pet plum tree and look at it. Its beauty exceeds even that of a laden apple tree. The shadows of the branches are flung across the rounded shapes of the plums; sticky drops of plum juice ooze out of some of the fruit. Others have a dust of bloom on them. A few bluebottles and wasps are making holes in the ripest plums, clearing them out and leaving the empty skin in perfect form and appearance. The fruit sings in colour against the particular clear blue sky of early autumn, and the sun glistens on the leaves, or turns them yellow as one looks at them against the light. As the wasps excavate the plums, I laugh at the remembrance of dear old Aunt Sarah; she was determined that the plums should not be eaten by bird or insect, until I should be able to visit her and have my feast of them. So elaborately she would fasten round each fruit a little paper bag such as children buy their sweets in, and the tree would stand with its burden of white paper bags, to the amusement of the entire village. As I lie in the grass, there is no sound but the buzz of the insects in the plums and the chirping of many grasshoppers around me. One of them jumps upon my bare arm, rubbing his legs. He is a beautifully shaped creature, but dull of colour compared with the grasshoppers I have seen in France, their wings backed with bright red or blue. But his continuous chirping adds to the feeling of benign heat all about me.

One of the first things we do this month is to plant our new colchicum bulbs. Already the square of uncut grass in the middle of the front lawn is rich with them, but we are enlarging the area under bulbs and have some specially fine new ones to put in. These autumn crocus bulbs, as we all call them, have the strangest shapes imaginable. Below the ordinary base of the bulb the large white

growth of the new flower extends downwards, making the planting of them exceedingly difficult; we have, in fact, to dig especially shaped holes to suit their individual eccentricities. But it is exciting work, and we shall have to wait only a few weeks before the first flowers appear.

This gentle weather of alternating sun and rain is kind to our seeds. The boxes grow greener and thicker as wallflower and columbine, mixed foxgloves and violas strengthen and expand. This year we are trying alpine strawberries, to be used as borders, and we watch their box with especial excitement. Gardeners get a double supply of pleasure, for always, while they are enjoying the actuality of the present and its blossoming, in imagination they enjoy in their planning the flowering of future plants. As we look at the small shoots of these new alpine strawberries we watch the berried borders of next June.

But in the autumn flowering of our garden one plant is missing: the dahlia. I visit other gardens, and see this flower in its amazing variety of form and colour, and I reproach myself for my stupidity. My mother had brought me up to have a terror of earwigs, and would never allow a dahlia in the garden or the house because it especially harboured them; and this grotesque taboo was stamped into me, and persisted long after I grew to like and respect the earwig. Only now have I emerged from its hold and will allow Noel to grow dahlias. Next year we shall have them of all sorts and colours. We may even grow as enthusiastic about them as a friend of ours whose entire summer holiday abroad was spoiled for him by the continual remembrance of the fact that he was missing his dahlias.

But at what time of year should we not feel that we were missing something by our absence? If it is not dahlias, it is peas or crocuses. The possession of a garden is an exacting tie. Someone told me the other day that nobody who was still young would consent to be dominated by a garden, and perhaps he was right. Perhaps that is why, as the blacksmith tells us, half the village allotments are untilled; the young men will not have their leisure controlled by the earth, for she is a relentless mistress. To us the excitement of taming the earth seems worth this tie. In a world where science shelters us from all the hardness of life, gardening gives us our only chance of a stimulating battle with the elements.

The bounding hedges live their own life. Throughout the year they are rich in their changing raiment, whether it be privet bloom or spindle berry, elder flower or wild rose. In the garden of the hedges

grow the many flowers that love the darkness, like the cuckoo-pint, hiding in the gloom of the hedge bottom. Over the body of the hedges, grateful for support, twist the lovely bryonies, black and white. The ditches grow dim with cow parsley and hemlock, yellow with charlock, rust red with sorrel; for if we protect our flower beds from these weeds there is no reason why we should not allow them to grow in our hedges. That is a good reason for a hedge garden. We

have had stubborn fights with gardeners for the life of our bryonies and our old man's beard. Each season when the time for the cutting of the hedges comes round, we have actually to stand and forbid the gardeners to destroy them. They tell us tales of the destruction worked on hedges especially by old man's beard, but we are firm. For a hedge would have to be a wonder of beauty if it should be lovelier than these creeping things. Noel has at last erected a wooden framework over which the old man's beard may fling itself at will; nobody can say we are murdering our hedges. In the hedges live caterpillars and slugs. It is to feast on them that our neighbour's tortoise is often to be seen creeping through gaps in the ditch, to crawl as far as his restraining string will allow him. Above all, the hedges are the homes of our birds. In their darkness they are secure. Among their thick branches they can build their nests. In these days when the countryside is being tidied, and hedgerows everywhere are being destroyed, birds must have lost many thick places for their nests. Let us see that we do not deprive them of their homes.

As the year has its round of flowers and weeds, fruit and vegetables, so does it have its cycle of insects. Now in the heat of late summer, the garden is blurred with winged ants, equipped for their nuptial flight. Earwigs abound, true to the calendar, the young they have so faithfully cherished now full grown. For the earwig is a proverbially good mother. We find them clustered everywhere, in the broken cleft of a tree trunk, in the shade of the bottom of the tits' hopper. There is an epidemic of them in this dry weather, and I am glad that I have recovered from my childhood's fears, and like them.

As we gather plums and loganberries, we look for the coming of the wasps. But this year there are few, and we miss them with their beautiful colouring and elegant shape. For there has been a campaign for the killing of the queens. Annie is glad, for she and Noel always have over wasps the only tussle of the year. Annie, for all her kindness, is for killing them; Noel rebukes her and will not allow it. This year there is rejoicing in the kitchen. Where usually it swarmed with wasps there are now but about ten.

The morning walk round the garden grows exciting, for windfalls are beginning. After a night of gale we pick up a basketful of apples and pears, damsons and plums. Each day now when we visit the orchard there is added to the beauty of it all, with the reddening apples and the bloom on the plums, the conjecture as to how many windfalls have dropped since the day before.

But there is one windfall that saddens us, and that is our mulberries. It is the first fruit we have ever had off the little trees, and we have watched them throughout the summer with great care and affection. Now they are dropping, and we manage only to pick one or two off the trees themselves. And they are sour, parched little fruits, interesting only in that we feel proud of having procured any of them at all.

Today as I sit in the house I hear a flutter of traffic outside the window on the tits' hopper. And there I see eight tits, excitedly trying the new nuts. For I have put out some cashews for them, and apparently they are very popular. The great tits are the greediest, and one enormous creature sits there and eats for fully a quarter of an hour, chasing away the more timid little coal tits and the blues. How nice it is to have them all back here again in our garden. The blue tits disappeared at nesting time and were away for months. We miss them more than we should miss any other birds.

Seasonal movements occur even among birds that are faithful to our own climate and spend their time in a restricted territory, though they are less obvious than the great dramatic journeys of migratory birds. The cuckoo and the swallow blazen forth their arrival and departure, but the little tits have their time of absence from the garden, when they go in a mass to the pine woods to feast on some special delicacy that grows only there. And even the London sparrows take an annual bank holiday at harvest time, arriving in shoals on the hedgerows bordering the harvest fields, to help the farmer with his gleaning. Chaffinches go off for a long stag party early each autumn, and Noel tells me a nice story of an excursion he watched. He was crossing to Ireland at the end of September and noticed about twenty cock chaffinches flying beside the boat the whole time, resting occasionally on the rigging and allowing themselves to be fed, but otherwise taking no notice of the ship whatever. They followed the boat with their regular dipping flight, like the shape of festoons hung across a street, until the Irish coast came into sight. A similar boat was starting back for England, and the whole stag party of chaffinches left Noel's ship and attached themselves to the returning boat. Very obviously they were out for a spree, unaccompanied by their wives.

The tits are flying around the hopper the whole morning, twittering with impatience to reach the new nuts and quarrelling with the particularly greedy ones who overstay their turn. It is a morning of birds. Swallows are sweeping low in the air. If they were to leave a line, like a spider's thread, behind them in their flight, what patterns would be drawn upon the sky. Linnets sing on the telegraph wires, and though it may not be the time of year for ecstasy in their song, just now all the birds are very friendly.

The robins have returned to us from their summer stay in the woods. In the general silence of birds we get bracing pleasure from their little crackling song. First came a young hen, speckled and brown, looking like a very small thrush. As we watched her we wondered how long she would hop around the garden mateless, and today her cock has arrived, the first signs of red bursting upon his breast.

Annie announces the start of autumn. As she arrives she tells us that it is "real finger-cold". No more for this year now shall we be told that there is thunder about. This phrase of hers is as seasonable as the mists of early morning or the reddening of the apples in the orchard.

And when we go into the garden this morning we find the first autumn crocuses in bloom. There, in the middle of the lawn, damp with dew, stand three pale mauve flowers, slowly opening their petals to the genial autumn sun.

SEPTEMBER

"LAST NIGHT'S THUNDERSTORM," says Annie, as she comes into the kitchen with an apronful of windfalls. She tumbles the apples out upon the table: green, yellow and red, cold and wet from lying in the long grass below the trees, some of them stickily sweaty where the morning sun has warmed them. Each day now we pick up windfalls, and yet the orchard still shines and glows with red apples. Have they ever before been so red? we ask ourselves. So have we asked ourselves each September since the young trees were planted and changed wild field into orchard. As we walk down the garden, in an arcade of apple trees, the gathering fever grips me. I touch the reddest apples, certain that it is their moment for harvest. But Noel holds me back.

"Not until the fruit comes off in your hand when you give it the very slightest tilt upwards," he says pompously, trying to hide the fact that he, too, burns to pick.

And there is no doubt that the apples do not yet come off easily when they are tilted upwards. But when Noel is looking the other way I manage accidentally to give one a slight twist as I tilt it, and it is in my hand, heavy and sticky. And when I am not looking, he, too, must have given one a slight twist, for he turns to me with an apple in his hand.

"It came off."

"So did mine."

But we both know better. And the apple gathering waits, though the fruit shines dark against blue sky.

Autumn begins now truly to clutch at the garden, gently showing the marks of its fingers in heavy morning dews. It is interesting to notice how differently plants absorb the dew; long after the flower beds and the hedges have dried in the sun, the mulleins will be silver still with moisture which catches and rests in the rough blankety texture of their leaves. And as we look across the garden at these lines and spots of silver wool, we are aghast at the number of mullein seedlings that have sown themselves. For it is so short a time since our garden was rough meadowland, and then later large naked beds of chalk, that we cannot yet understand superabundance. I remember the appetites of those empty flower beds, as they swallowed up bulbs and plants and seeds and bushes; and still there would be gaps to shame us, and weeks of space between fadings and flowerings. Neighbours were kind and would give us their overflow. Annie brought us cuttings and slips. Nurserymen's catalogues cast spells over us, till we squandered our savings on anything that would fill those empty spaces. And now, after four years of rain and drought, withering wind and clamping frost, the rough meadowland begins to look like a mellow garden.

It is a great moment when a garden becomes bountiful. And it is another great moment when, for the first time, one can give away plants.

But there remains the problem of the mullein seedlings.

Already we have found good homes for scores of them. One plans for them as though they were kittens. When neighbours pass down the lane I pounce upon them.

"Wouldn't you like some beautiful mullein seedlings?" I plead.

Mrs Burr is acquisitive and goes home with both hands full of healthy young plants. Mrs Headley has always had a passion for mulleins, and hurries back to fill all her odd corners with them. Mr Gomme is a social snob about his garden and considers the mullein a common plant. Little Miss Carter does not know a mullein by sight, but is glad of anything to fill her new beds. And so we realise with a guilty feeling that in a year or two the whole county of Bucks will be one solid mullein plantation. But still seedlings remain. This summer's sun has put great force into the old plants that they should spill their seed so generously over the land. Though we are sentimental about them, it is no good. For the first time we are going to have to destroy. Miserably we dig them up and bury them, realising that now that abundance is upon us, we despise it. It is little consolation that the decaying plants will make much needed humus.

As we look round, we see signs everywhere of the same riotous abundance of self-sown seedlings. This hot summer has stimulated the body of the garden and each plant has fulfilled its purpose to the uttermost. But how have the seeds drifted so far from the mother plants? Is it the bees, or the wind, or have we perhaps carried them on our feet as we walked? For foxgloves dot the rose-bed, hollyhocks break into the rows of lettuces, and violas are at the far end of the garden among the black-currants. We try to persuade ourselves that we enjoy surprises, but we are secretly relieved when one morning we see that Darville has destroyed these trespassers.

How gentle this September sun is, and kind. No longer is there cruelty in it at midday, so that one runs from its blaze into coolness;

rather does one grasp at its warmth, knowing that so soon it will be gone. The whole garden of flowers glows under it, with a misty golden colour that the glaring summer lacked. From the changing colour of light in the different seasons I believe I could tell the time of year if I were to see neither trees nor flowers nor birds, but only

light. Is it the angle of the sun that causes this, or the amount of moisture in the air, or what?

In this golden glow the perennial bed blooms with Michaelmas daisies and chrysanthemums, asters, violas and giant balsams. It has never been more beautiful. No longer do we shrink from looking at our plants as we did during the summer's drought, for the rains have swelled the leaves once more, the flowers are happy in their second blooming and the grass at last is really green.

I mow the lawn. How many people know the right way it should be done? Feet should be bare; grass should be slightly damp. The cold, moist clover strikes up from the mower upon my bare feet, and blades of cut grass and bits of slashed weeds stick between my toes. I remember one of my moments in life: I had been bicycling among the foothills of the Pyrenees at the time of the grape harvest, and, turning a corner had come upon a shouting clump of peasants. They were laughing as they sang odd little songs, and some of them were dancing. In the midst of them, and all among them, and surrounding them, were squat fat tubs, some full of grapes, others purple pools of juice. And in others again were men and women treading the grapes. They beckoned to me, and I, too, trod with bare feet. The grape stalks and the pressed skins were caught between my toes, and as I drew my feet up, there was a resisting, sucking noise, as if I were pulling against a tide. It was a joy I had never known. I was outside of time. So had I trodden grapes centuries back. And the stalks and the pressed skins turned green and became grass and clover, and I am treading the lawn with little jerks from the machine as it comes against a trespassing dandelion or a dock. Suddenly I feel a sharp, bracing prick on my foot from a young thistle that also should not be there, and the stinging hurt shivers up my body. But the whole time there is this feeling of goodness coming up into me from the ground. We are losing much, these days, when we no longer get this

naked contact with the earth. The sensation of touch seems to be fading, and lazily we look at things with our eyes, and smell the more pronounced scents around us, ignoring the vast range of emotion that is within the scope of hand or foot. Few think of caressing a flower and enjoying the feel of its form and texture, the tightness of its bud, the hardness of its seed pod, or know the pleasure the hand can get from the surface of a tree trunk or a vegetable marrow. The peasant in the field will see a small bird afar off, or smell the change in the weather, or get happiness from the feel of his soil. We are poor creatures that we should call ourselves civilised, we who have only these blunted powers.

The sun that shines through the mauve, check-patterned petals of the autumn crocuses on the front lawn ripens our tomatoes. This is the pride of our year. To have grown a row of tomato plants in the open, one with forty-eight fruit tumbling heavily down it, and the rest with nearly as many! They are so beautiful in their dripping clusters, with their musty acrid smell. It will be a race now between the September sun ripening them and the autumn frosts smiting them. Already I take off those fruits that have turned a dull yellow, that they may redden in the sheltered sun in our windows. It is so difficult to find the dividing line in one's garden between utility and beauty. If things are eatable they are supposed to be only useful, if they are flowers they must be merely decorative. But our tomatoes are lovelier than most flowers, and if we have to tell the truth we must say that we only grow vegetable marrows and scarlet runners for the beauty of their blooms. To eat they are dull, but to look at they are disturbingly lovely. On the other hand, beautiful though sunflowers are, I doubt if we should always aim at having them in our garden if it were not to provide, with their ripened seeds, an autumn delicacy for the tits.

These days of sun demoralise us. We should be working and gardening hard. But it is lovely to lie on my back under the chestnut tree, looking up at the sky; to feel patches of warmth moving over my body as the sun shifts behind different branches and leaves of the trees; to see the insects hover above me and to wonder if, with luck, just one of the swallows that twitter high in the air will sweep low near my hand to catch the insects. But for some time now the sun has been behind the thicker part of the chestnut leaves. I turn over and lie on my stomach and watch the world of moving, living things beneath me in the grass. I become a Lilliputian and meet ants; here is one carrying the dead body of a comrade through a forest of blades of grass; another milks a green fly on a speedwell stalk; yet others climb with effort over bits of chalk and sharp-edged stones, up hills and down valleys, as they remove to safety their white, bladder-like eggs. Shiny pink and brown worms squeeze themselves up from the earth and coil among the grasses, their veins showing through their transparent bodies, their girth broadening and lessening as they move; they pass little red pimpernels and daisies. On these little flowers flies and bees alight and feed. Here are minute hopping creatures and earwigs and small beetles and very little spiders, each living its own life and unaware of me. Everywhere around, sticking up from the ground, are the discarded pupa cases of daddy-longlegs. I watch a daddy-longlegs as it helps another out from its case, seeming to rip open the tight-fitting imprisoning jacket. I wonder

if this is really what it is doing. I watch others mating. The grass moves with them, for the daddy-longlegs is a plague this year. He is a friendly creature, but his larva, the leather jacket, feeds on the roots of the grass. Next year, then, we shall suffer. Annie tells us it is all the fault of the people around here who upset the balance of nature with their campaign for the killing of the queen wasps, for the wasps eat the eggs of the daddy-longlegs. It seems as if she is right when we recollect that this year we have hardly seen a wasp. The ground quivers with coupled daddy-longlegs, which also catch in our hair, our clothes and our feet as we walk in the garden at dusk.

As dusk rolls into night we watch the stars. They are bright as they were not in the summer, for the sting of cold in the autumn night air would seem to make them sparkle. And now, as we look upwards, they lodge in the topmost branches of the trees like silver fruit, to fling themselves, rocket-like, across the sky. It is the season of shooting stars, and they riot over the ink-coloured darkness like balls in a quick game. One carries a tail behind it. But, bright as they are, they do not lighten the blackness of the garden as we walk in it. It is all so formless that I cannot see the bats that shake in the air around me, but can only tell that they are there by the vibration as they pass near me, or by their tiny high-pitched cry. Noel grows exasperated, for he cannot hear the cry of the bats, and likes to pretend that I am imagining it. He says the pitch is known to be too high for some ears, and that I ought to think myself very lucky that I am able to hear it. I wonder how many sounds there are that none of us can hear, and whether we get the full song of birds. Noel

says that cats only hear high notes. He has whistled to them the same tune in various octaves; to the lower ones they will remain indifferent, but to the higher ones they will fling themselves on their backs and roll with delight. We have also noticed that the tits answer with twittering only to the lighter, higher pitched records on our gramophone. After much experimenting we have found that they have an especial favourite in Mozart's Oboe Quartet. Perhaps they catch a likeness in the sound of the oboe to their own notes, for they seem to try almost to drown it with their song.

But now we hear the tweet of the little owl and creep to the bottom of the garden to see it. We must tread cautiously, for Cornelius, the hedgehog, creeps around at night, and several times we have nearly stamped upon him.

Then a disaster happens. Darville leaves us. No mother ever feared the loss of her baby's nurse more than we fear the loss of our gardener. He knows the temper and health of each of our garden's children. We introduce the new gardener to the place, and show him to each plant and tree. But he has not got an earthy look, and we feel unhappy. What a real thing this earthy look is. Alf, the woodman, has it. Is it some rhythmic sympathy with the earth that shows itself even in the way clothes are worn? It is not the actual colour of the dress, for to be one with our pale chalk a man would have to wear a garment of light grey; and it is not the shagginess of the person, either. It must come from the attitude towards the earth. The new gardener looks upon it as something that earns him money for bread, while Alf is at one with it in its moods and substance, himself as rough as the ground he digs.

The grip of autumn tightens and the end of the month brings cold and wet. Golden sunsets and shafts of light on the lawn are seen only in memory. The rain comes down in a grey mist, sending us shivering into the house. Darkness falls early, and we begin to pity the late blooming roses left out in the cold all night. Equinoctial gales rage. The storms snap off branches of apples from the younger trees, and the last of the autumn crocuses on the lawn are slashed and windblown; but how gently they lay their heads down flat on to the grass to die, making of death a gracious thing.

The autumn session of the starlings' parliament has been summoned. Hundreds of them chatter in the thinning branches of the two big elms outside in the lane. Their noise is so loud that it almost drowns the much nearer sound of the swallows that crowd on the telegraph wires: the tribe in which we have sojourned all the summer is striking its tents.

We, too, brace ourselves to do things. We visit the shed, and look at the garden tools, feeling along the edges of sickle and shears. They are blunt, and we take an armful to the village blacksmith, that they may be sharpened. For the garden now becomes our master and before us stretch weeks of autumn work, digging and planning, tree planting and manuring. We live here at the dictate of the weather. In the city summer merges into autumn and a man going to work suddenly sees that the leaves are falling from the trees; as autumn changes to winter, he notices that nightfall comes earlier, and the cold grows stronger. But to live vividly one should have vested interests in the weather. In the garden we have seen our plants and trees wilt through the burning months of drought; we have watched grass seed washed away by a sudden hailstorm; our wallflowers have been shrivelled in spring by the north-east wind. A persistent summer sun has browned our lawns, and a whole row of spinach has been blackened in a night by an early frost. Sun and wind, rain and frost, are powers that cannot be ignored. This same sun burns down upon the city roofs; this same rain soaks its pavements; but to us in our garden sun means the opening of buds and the ripening of fruit, while rain brings growth and life to our plants. Our senses feel each change of the weather and watch the great arc of the sky from horizon to horizon for signs. The keyboard of experience expands.

OCTOBER

OUR POTATO CROP waits to be dug. As I fork up a root and scrape a potato with my fingernail the skin slips off. All around us work shouts to be done. We have no time now for quiet enjoyment of our garden, for the first frosts and the autumn rains will soon be upon us, checking our digging and planting.

We enjoy digging our potatoes. It is the big treasure hunt of the year, even more exciting than searching for the fruit in the tangle of straw round the strawberry plants. The excitement lies in the anticipation we feel each time we stick the fork into the ground. How many potatoes will there be beneath this plant? This anticipation never tires, even after rows of digging. Here is all the mystery of an unknown, invisible harvest. We can see the extent of our peas and beans, and we know that each green-leafed parsnip top will have

a corresponding root below, but who can tell how many potatoes
huddle beneath the plant that we see above the ground? As my fork
brings up the cool, moist potatoes, I lay them out in the sun to dry.
They look beautiful as they lie on the earth in creamy rows. The
limp, fading haulms curve away from them by their side in regular
lines. Minute, undeveloped potatoes cling to the tendril roots of the
plants, smooth of skin and fresh of colour in contrast with the decay
of the aged seed potato. A robin sits near us on the haft of a spade
and sings his autumn song to the worms that I unearth; they are
many, and they wriggle back below the soil as fast as they can. But
the robin is too quick for most of them, and he has a great feast.
From time to time my fork spears a potato; in its damaged centre
I find lovely small pink worms coiled tightly round. Annie and
the gardener do not share my enthusiasm over the beauty of these
worms. They tell me that we should have dug our potatoes earlier, as
sensible people have done. The damaged potatoes will not keep.

Digging exposes an independent living world below the surface
of the earth. Among the persisting roots of bindweed is the home
of the burnt-coloured brittle wireworm; here it will live for nearly
six years before it escapes above the soil as the click beetle. The
little woodlouse wanders about, ready to coil itself into a ball at the
slightest vibration. The centipede twists its body until it is the shape
of a switchback at a fun fair. Rubble and bits of broken crockery
speak to us of men who knew this field before our day. But it is a
world that flings us back in time and makes of the clods we turn
pages of history, for one day among the marigolds we dug up a
local money token of the 18th century. Relics of Roman Britain are
scattered among our earth as broken pieces of brick; they are many,
for we are on the edge of the Icknield Way, which was used as a
Roman road. Our imagination one day was especially excited by
the finding of the tooth of a wild boar. Instantly our garden seemed

full of perils, and civilisation shrank to a thin veneer of three inches of chalky soil. We have kept the tooth, and when life grows too respectable and secure, a glance at it reassures us with a thrill of fear.

But we grow bored with the endless surgical work in the garden. The summer's drought had checked the growth of the weeds, but now, after the September rains, they burst above the earth and cover the garden with enviable exuberance.

We have decided that the only inspiriting way to rid our garden of weeds is to organize a campaign against special enemies. If we destroy certain weeds each year we may hope one day to have a clear garden. Unfortunately, we are surrounded on most sides by meadowland and unkempt gardens, so that dandelion seed and thistledown drift across our hedges in the wind and settle happily on our flower beds. It is a hard struggle. This year we have been

waging war against groundsel and bindweed. Our fighting spirits have been so strongly roused that I have known us instinctively get off our bicycles in the country lanes to destroy a particularly big clump of groundsel. The groundsel is an easy enemy to vanquish, for its root comes out of the ground with ease, and one has no qualms about its destruction. But the bindweed is so lovely, with its flowers striped like the pink and white cotton frocks of young girls, that one is tempted to say that it is not a weed. It was only when our plants were nearly strangled by it that we became ruthless. Now we search the earth for the loose white strings of its roots, knowing that it is a wandering and tenacious creature.

There is some hope in weeding, for the weeds may one day be defeated, but the tidying of the garden is as exacting and unending as the daily washing of dishes. Soon the time of growth will stop for the winter, but still the lawns grow against the edges of the flower beds, and still the grass banks need regular clipping.

It is a relief from the constraint of all this small work to turn to planning. This year we are making many changes. I find that I am very conservative in my garden. I like to keep the same flower beds, although I know that Noel's proposed alterations will be an improvement. There is something desolate and unfriendly even in the inevitable changes in the vegetable beds, as the clearing away of the dead broad beans exposes large expanses of earth, or the digging of the potatoes removes the thick rows of green haulms.

And now there is a large new bed to be made. We have imported some rich, brown "foreign" earth for it, and as we heap it on the bank we feel ashamed. Somehow, it seems like cheating, and it is an insult to the garden; but we are determined somewhere to have chalk-free soil, that we may grow those flowers that are now impossible. Our shame is still greater when new grass turves arrive. Here, surely, we are cheating. We should sow our grass, and not get it ready made. We try to justify ourselves by saying that it is too late in the year for grass seed to grow, but we feel guilty. The turves are beautiful and put our weed-choked lawns to shame; they are real grass, with a backing of genuine *brown* earth. Annie has a great reverence for them, since the carter told her they had come from some nobleman's park. "No wonder they're so good," she says, forgetting her usual classlessness. It is fascinating work to lay them and fit them in. "Stair-carpets!" says Annie, with a sneer, as she leans out of an upper window and sees the new gardener helping us with the turves; for she has a grudge against him, because he started life as an indoor servant. Her Alf is a true countryman. The fun of laying them increases as we finish the straight portions, and I cut out darts in the turves and long angular pieces to insert between the slanting edges. It is with a feeling of power that I watch the face of the garden change under my hands and eyes. The shabby untidiness in front of me is curbed and civilised. In a few hours we have made a new grass bank.

But it is the planning of the new trees that is our main pleasure. Noel has a consuming lust for it, and spends an afternoon in a stupor of emotion, measuring distances between proposed new trees, writing down the numbers required and adding them up. Sixty-six new poplars are needed, and they only cost fifty shillings a hundred! Trees are grotesquely cheap. It always amazes me to realise that one can buy a future landmark for a shilling, and make an avenue out of the cost of a new dress. But it is wrong, anyhow, to think of trees in terms of money. They should be beyond the world of ledgers, even as there is no charge for a white cloud or for the warmth of the sun. Noel is happy, for he manages to find many places into which he can put all these new trees he means to buy. Having dealt with the poplars, he decides that we need two elms, for we have none and they are beautiful. Chestnuts come next into his mind, and he realises sadly that we have space for only two; this will not do. Chestnuts are lovely trees to plant. He thinks again, and suddenly he has an idea. We can have a row of chestnuts in our field across the lane. Better still, we can also give some to our neighbours, to hide their houses from us and to make the whole place around us look green and beautiful in the summer. He rushes out of the gate to the neighbouring houses, offering them presents of trees. Within a few minutes he has found willing acceptances and he is happy, for his list swells. Finally we come to the fruit trees. We decide that we can squeeze two more apple trees into the front orchard. They should be cookers and they must be bright green. We bring out catalogues of fruit trees, and the afternoon fades into evening, and dusk sends us into the lighted house before we have been able to pull ourselves away from the lists of apple trees.

It is no wonder that we spend so much time with the fruit tree catalogues, for they contain the most engaging descriptions. In them we learn that one apple tree is a "shy bearer." Another "is not suitable for orchards where cattle graze, as it has a weeping habit." Yet another "must be eaten at the critical moment," for it keeps after its flavour has gone, and we may put our teeth into it a few moments too late and miss its essence. The Irish Peach should be eaten from the tree. This would seem to make of apple eating a most serious occupation. We wonder if we have hitherto been too casual. The life histories of the fruit are exposed in these pages, differing from a social register only in the indiscreet manner in which the family skeletons are dragged from their cupboards. We read that Allen's Everlasting is "a bad character," and it is with disappointment that we learn that he is so condemned merely because he does not ripen in cold years. Our malice whetted, we see that Beauty of Bath "is self-sterile," a condition not uncommon in that resort, and that Baumann's Reinette is "more pleasing to the eye than to the palate."

Another lady is "subject to scab," and several are "apt to canker." Very few emerge with their characters unspotted. But, in spite of these home truths, we are given pictures of their beauty, with their red-flushed cheeks and their highly flavoured pale yellow flesh. Their pedigrees are long and reputable, though, to be sure, Gladstone was obscurely found in a field by a Mr Jackson half way through the last century. It is interesting to see that the apple world lacks the social snobbishness of the rose world. There may be vague Emperors and Queens, and a few lords, but for the most part we meet plain Annie Elizabeth, or Clark's Seedling, or Sturmer Pippin.

Yes, it takes a long time to make out one's lists of new trees.

While we are planning like this, we must not forget to look about us and enjoy the autumn in our garden. Have ever the hedges so flamed with berries? Scarlet rose hip and bryony, bright like coloured beads, outshine all else, dulling the heavy deep red of the hawthorn fruit. Hanging sprays of ripe blackberry sparkle against overweighted clumps of black elderberry; the traveller's joy has not yet turned to smoke. Annie picks the blackberries as fast as she can, for jelly, before the Devil shall have spat on them at the middle of the month. Blackbirds and thrushes are as busy with the elderberries. Noel wants to trim the hedges, but feels that he has no right to cut away the birds' larder, and we decide to leave it until after everything has been eaten.

But it is not only the blackberries and the elderberries that are being harvested. Everything that can now be gathered must be brought into shelter against coming frosts. We strip the last tomatoes from their plants, sadly missing their bright spots of colour among the vegetable beds; hazel nuts from the hedges must be picked before they all fall to the ground. We gather them one evening at dusk, when we can hardly see them in their protective colouring. The branches rustle as we pull them down and search for the nuts. But I gather together the

fallen hazels, thinking to outdo Noel in the quantity I have heaped
into my basket, only to find that he is right when he says they will
be the unsound ones that have been eaten by maggots. We eat some
of our harvest the same evening, enjoying the white of the inside of
the freshly gathered nuts.

This is the time of year when everything drops into the earth. In
spring there is an upward movement all around one, with a lift in
plants and trees. Now it is the time of weight, when seed pod and
berry, fruit and leaf fall and return to the earth. It is truly the Fall,
a lovelier word for this season than autumn. The horse chestnut
has cast down its shining fruit, warm of colour as it breaks from
its tight-fitting, kid-lined case; on the soggy ground all round lie
these lumpy, horned shells. The winds blow down pears, and we
find them, yellow and brown, surrounding the trees at their base.
So we strip the fruit trees, our pleasure in picking tempered by the
sight of the trees bare of colour for another year. It will be such fun
when our young apple trees have grown tall enough to need ladders
for the apple gathering. For this is an essential part of the ritual. I
remember Aunt Sarah's trees in Berkshire when I was a child. They
were old and gnarled and high, and I climbed ladders and strained
and stretched to reach the topmost bough. Always the highest apple
would shake away from me and stay there in defiance at the tip of
the tree well into early winter, until one night of storm would blow
it to the ground, where it would lie bruised and weather-worn, to be
eaten by ants. But this will come with years.

As the garden gives up its fruit, so must we feed it, to renew its
strength. Realist though I am, I find it hard to grow enthusiastic
about the close-up smell of pig manure. One may pass a farmyard
and enjoy the diffused smell of "muck" that drifts out, but it is a
different thing when the pile of manure is outside one's front door.
The carter from the farm down in the plain has dumped our manure

literally a few yards from the door. It is grand stuff, short and with very little straw; but that makes it smell even stronger. Noel and I set to work to move it. The gardener is away ill, but it is only right that we should ourselves do this, if we are to enjoy the flowers that will benefit from it. Fortunately, we grow more accustomed to the smell as we work, though, as we distribute it over all the garden there is no refuge in which we can escape it. The roses have bloomed this year in such abundance, and there has been so little nourishment in the dry soil, that we must be liberal as we feed the garden. We console ourselves for the unpleasantness of the work by telling each other that our gratitude to the flowers and trees is a poor thing if it shrinks from an unpleasant smell. As we cart barrow-loads of the food to different parts of the rose bed and the vegetable plots, I mind the smell less by recollecting the glamour that manure always holds for gardeners. I have never heard one of them mention the smell, except approvingly. I think they feel that the stronger the smell the better the stuff. Especially I remember our first gardener, Anderson. He used to lick his lips and his eyes would gleam and glow as he saw a load of steaming pig manure. "That's good stuff," he would purr. "That's lovely, that's lovely. And don't it smell grandly ripe! Beautiful stuff!" With delight he would toss it down over all the beds, and dig it in round a tree, stamping on the bare manure in his exuberance. He was the true romantic. And so why should I allow an ecstasy to pass me by? It's grand stuff, this manure!

As we feed the garden, we review the damage done by the summer's drought. In persistent rain we look at the dead beech bushes that would have been saved by even a fraction of the wet we are having now. We count the gaps: four in the front, six at the back and one against the rubbish heap. Two or three cherry trees are half dead, but may just be saved. Several of the clematis plants are still unhappy. While we look about us, Noel decides to alter the plan of the garden. If we move the manure heap we can make a path that will give us an uninterrupted walk right round the garden. This, luckily, will mean buying more beech trees for a new hedge. Back into Noel's face comes the intoxicated glow that I know so well. And more trees are added to his list. This new plan will also mean the moving of many trees. Next best to buying trees, Noel enjoys moving them. So it is with satisfaction that he realises that we shall have to move two young cedars, his pet walnut tree, one mulberry and the whole top line of fruit trees.

How it rains these days. In the grass by the hedges, large shiny slugs appear, black as liquorice and beautiful of shape as they stretch themselves out. They heave like ships on a rough sea in their passage across the grasses. The garden is sodden and the trees drip, their autumn colours deepened and burnished by the wet. But roses and violas still bloom, and carnations are in bursting bud. Michaelmas daisies are untouched by frost and the cosmos still shows pink among its seeding heads. We can gather bowls of bright-coloured flowers for the house.

Into this sodden, dismal wet strike gleams of sun, with rainbows. In one day I see six. They link up hill with hill, bewitching the beechwoods that already turn colour for the Fall.

It is the month of the Hunter's Moon. We walk in the garden in moonlight. The rows of leeks and celery and parsnips assume a look of importance, mysterious in their masses of dark shade and silver flecks. The trees throw shadows across the ground, demonstrating the subtleties of their shape to us as they can never do in the daytime. Each small bush adds grandeur to its person. This silver light ennobles everything.

And then, on a dismal afternoon at the end of the month, I hear a particularly sweet bird song. Outside, in one of the flower beds, is a goldfinch on a cosmos plant. He has discovered that the cosmos is in seed. As I am enjoying the beauty of his red and black and gold, he flies off, leaving the flowerbed colourless and dull. I am turning from the window, disappointed at the shortness of his visit, when I hear a rush of wings and look out again at the flower bed. The pioneer goldfinch has returned, bringing with him a "charm" of the gaily coloured birds. They sing and twitter as they fly from plant to plant, pecking at the ripened seeds. I stand quite near them at the open window and they are unafraid. In their brilliant colours and their drooping flight they are like lighted Chinese lanterns swinging in a garden in a slight breeze. They have illumined for us a grey day.

In this time of decay in the garden the rhythm of life never stops. The chestnut tree throws off its leaves, exposing, where the leaf stem had been attached, that shapely mark of a horseshoe, nails and all, that gives to the tree its name; but it is the swelling of next year's bud that pushes off the old leaf. Around the dead stalks of this year's lilies sprout green shoots of next year's plants. Growth has started in the foxglove seedlings, and young cornflowers appear around the still flowering plants. Autumn is not the sad time it is supposed to be. Darkness falls at five o'clock, and the garden is cold and wet, but it is a season of planning and expectation. It is now that we plant our bulbs, in itself an act of faith. How, then, can autumn be called dull and hopeless? Even the fallen leaf is food for future years of foliage and fruit, and promises next summer an added colour to the flowers.

NOVEMBER

"WE'VE HAD ICE the thickness of a penny," says Alf as he comes up to help with the digging.

It is the morning after the first frost, and fearfully we go into the garden to look at our plants. Great damage has been done all round, for flowers have bloomed strongly in the warm, damp weather of October. This is no gradual, beautiful death, as is their normal due, but a cruelly sudden ending, with blackened stalks and leaves and the distortion of sharp decay. Flowers die in such different ways. Some, like the iris, wither and grow ugly on the stem. Most of the spring flowers, such as daffodil and hyacinth, fade and brown and crinkle, till one cuts them down as they pass the height of their blooming, rather than behold them in their old age. The poppy is hasty and dramatic. The kindest flowers are those that cast their petals on the ground, and in dying lose none of their beauty. Tulips die like this, and roses. I remember, too, the death of our columbines last June; as the seed pods swelled they thrust off the purple petals, scattering them in the wind over the grass, till the stems supported nothing but austerely graceful seed pods. There was no moment of this death in which they were not beautiful. But now snapdragon and cosmos hang blackened and miserable, and Michaelmas daisies are rusted. The garden this morning is not a happy place.

Perhaps the saddest sight of all is our Indian convolvulus. Throughout the extreme heat of this summer we have watched its growth and wondered whether it would bloom before the power of the sun should lessen. Noel had seen it in Sind, a bountiful, blue creature that clambered high over walls and doorways. Seeds of it were given to him, and with much misgiving we had planted them. For how could we hope to coax into flower here in England a plant that lived in the heat of the Indian sun? But green shoots had appeared and sprouted, and tendrils had climbed up the south wall of our house. And then one day we noticed that the plants were studded with buds. A close race was beginning, for the season of frosts was near. Frightened of these early frosts, we had brought into the house the biggest bud with its pink tip turning to blue; but no indoor warmth would persuade it to grow, and for a week it has not developed. Now, in one night, the plants have been blackened and the buds destroyed.

We dig the holes for the new trees, that the frost may break up the clods of earth. Alf digs with us, a rhythmic figure as he sways and moves in his pink and blue striped shirt and loose brown corduroy trousers. His massive boots have iron tips to them. He reminds me of an upright brown hairy caterpillar. He spits on his hand before grasping the spade, by habit using all the little tricks of his craft. The solid chalk below the earth is so hard that we have to break it up with an iron bar. How can anything grow in this earth of chunks of chalk? As we look across the orchard to the fruit trees that have been fed round their roots with dark circles of manure, we imagine how lovely our garden would look if only it had dark soil instead of this chalk. We dig, and the garden becomes a pattern of round grey holes, awaiting trees.

But apart from the sudden wreckage of the frost, there is now a slowing down of the processes of the body of the garden. Leave the garden unvisited even for a day in April or May and plants will have rushed up visibly, and buds will have appeared; but now there is little difference in things over a week or two. Only more leaves fall, so that the bony structure of the trees grows increasingly evident, and plants gradually die down in the flower beds and expose larger spaces of bare earth.

One thing alone changes rapidly at this time of the year, and that is the colour of the leaves on the trees. They are as bright as flowers, and replace in gaiety the fruit we have gathered from the garden. But even this change is slow as compared with the rush of spring, when the leaves burst from their buds and clothe the trees in a flash of time. Since the beginning of September the leaves have been turning, varying in date with tree and weather. Weeks of calm will keep the leaves on the trees, as though they were forgetting to fall. They will be torn off in one night of wind. All the leaves have by now returned their sap to the tree, except the undramatic walnut which, like the ash, refuses to end its year in a blaze of colour, and lingers until the frosts strip off its leaves and scatter them at its base in a dirty, blackish green heap. The garden now is riotous with colour, in gradations of yellow and gold, bronze and crimson. We are reconciled at this moment to our smoke-grey soil, for it is an unequalled foil to these flaming leaves.

Our beech hedges surpass in beauty even the cherry tree or the pear. They thrive in the chalk earth of our garden and are lovely the

whole year round, whether it be in the late spring, as bud opens to pleated leaf, or in the depth of winter when the leaves are crinkled and brown and crackle in the breeze with the sound as of thin metal shavings. For the beech hedges are never bare of leaf. Now, in November, they burn bronze and yellow and orange; their leaves are still soft with the wet, and do not yet crackle in the wind. Soon these soft leaves will shrink and dry and remain on the hedges until next April, to make a hiding place for the thrush's nest. Only when the new buds swell and grow insistent will these old brown leaves be pushed off.

Yellowest of all are the leaves of the plum; but they fell early, and the trees now are bare except for a few stray leaves that were out of reach of the winds in the shelter of a branch. The green at the tips of the poplars has turned to the colour of the lower leaves and now the whole avenues of poplars are golden. But soon these leaves, too, drop, and as the breeze touches the thinned trees it makes a noise like the falling of gentle rain. Each tree has its own characteristic sound in the wind as its foliage grows brittle and old. By the middle of the month most of the trees will be stripped bare and the ground below each tree will be covered with leaves, in crumpled flat heaps of colour, like the petticoat out of which a woman has stepped at bedtime: silver white under the whitebeam, burnt brown under the chestnut, ruddy and crimson under the pear. As we lose the colour from our trees, so do we grow to understand and love their form.

The hedges, too, are thin. At this time of year the birds brush off the fraily balanced leaves from the bushes as they move about. A blackbird shakes the orange seed from the pink cup of the spindle in his flustered flight from a cat. And as the leaves cover the ground, they make a protecting quilt for numberless hibernating insects and for the cocoons of next summer's butterflies. They shelter from cold and wind frail plants that will flower next spring. In a year or two these same leaves will have lost their shape and identity and will pass down into the earth to be transformed into food for the tree from which they came.

But the garden is full just now of rubbish that must be burnt, weeds that we dare not dig in, lest they should again take root or seed, prunings from the hedges, annuals killed by the frosts. Round the flower beds and the vegetable plots we go, collecting all we dare, with a good conscience, lay our hands upon. For who would not seek to have as large a bonfire as possible? With glee I cut down the giant sunflowers, removing their brown heads of ripened seeds for the tits, to be suspended by a string outside the windows; but the nine-foot high sunflower stalks and leaves are grand bonfire fodder. Alf is anxious to add all the pea-sticks. "You'd never get no peas to climb up a last year's pea-stick," he says. "Why, if some of the sticks be this year's and some be last year's, they'll make for the new ones and leave the old 'uns all bare. Beans, now, don't mind." But then,

Alf is the local woodman, and it is he who supplies our pea-sticks. So we risk the fastidious discrimination of our peas, and the sticks are not used to kindle our bonfire.

We light the bonfire. The fire catches in the heart of it, the smoke comes up in slow, white curls from the crater-like top. The dry sticks crackle, the earth round the roots of weeds damps down the flames. As it sparks and glows and smoulders, we are filled with some primitive feeling of fire-making. It is not just a grubby pile of dead groundsel and frost-bitten snapdragons that we watch; these swaying clouds of smoke are not consuming mere overgrown branches from a hawthorn hedge: we are in the presence of ageless fire, that we have made, and can control. It is with difficulty that we check ourselves from dancing round it. The bonfire simmers away the entire day, and with the stillness of dusk the white smoke spreads out low and flat among the hedges and across fields far beyond our garden. When night comes I go out and rekindle it, that we may see the flames

against the dark. It is good to have an excuse to light a bonfire.

As we search among the vegetable plots for fuel, I pause at the Brussels sprouts. They are beautiful with their tight fists of sprouts bunched up the stalks, and the long-stemmed leaves curving down all round the plants. I wonder why so few people remark their beauty. They look like a sculptor's work. The vegetables mature as the flower garden withers. Purple sprouting broccoli thickens and throws out its rich, blue-green foliage; banked rows of celery thrive in the frosts, each plant encased, for blanching, in newspaper and string. Leeks stand upright in their severe carved folds, rushing up to flower before we are ready to pull them. It is great fun to let

some vegetables run to flower. This summer, in an odd corner of the
garden, I cherished a blossoming leek; town-bred visitors gazed at
it, unable to guess what it was; country people said it was a waste
of a good vegetable: but it was a lovely object for weeks. In the heat
and the drought several lettuces in the salad bed sprang upwards
and flowered, throwing out their pale gold chandeliers of bloom.
Rhubarb in blossom has a rich, exotic beauty, matching in character
the flamboyancy of its foliage. The yellow cabbage flowers are as
lovely as cultivated blossoms, lighting up the vegetable plots in May.
Darville wanted always to cut them down, and was amazed at our
wishing him to leave them until they had finished flowering.

There is so much to be done in the garden just now. We wonder if
we shall manage it before the winter frosts clamp down the earth,
making all work impossible. The vegetable beds must be dug and
manured. The flower beds are still untended. Last summer we visited
a local nurseryman, selecting new rose trees from his blossoming
bushes; they were to be delivered in November and may arrive any
day now. We are late in planting our bulbs and must rush to put
them into the ground. Thanks to Noel and his new plan for the
garden, there are many trees to move. Finally, most important work
of all, there will soon be all the new trees to plant in the holes that
we have prepared for them. We decide first of all to move the trees.

Transplanting trees is agitating work. Recollecting the dislocation
and distress a human can feel at being transplanted into a new scene,
one cannot help worrying when one moves a tree. The nervous
shock to the tree, running up its trunk through its branches and
into the veins of every leaf from the disturbed roots and fibres, must
be extreme. This shock must be greater to a tree than to a human,
for we can reason and a tree cannot. I am unable to work for days
after removal to a new place. Is it any wonder that transplanted
trees so often die from shock? When the day comes that we can

give anaesthetics to our trees before subjecting them to the ordeal of removal, we shall feel much happier. Meanwhile, the mulberry must be transplanted.

We have prepared its new home. It is oblong and deep, to fit the shape of its twisted roots, which have dug themselves down into the solid chalk. So securely are they imprisoned in the chalk that it is hard work to unearth them. They do not look at all happy. We lay a bed of grass turves below the roots, that they may get more moisture. The rains would otherwise run over the chalk and leave

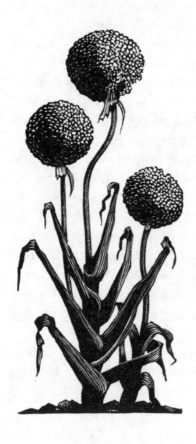

the roots dry and unnourished. I hold the mulberry tree upright in the hole, with its supporting stake, barely letting it touch the grass turves. Noel sifts the ashes of the bonfire all round the roots, feeling that he is helping it in its troubles. We cover it in with good earth, tuck up its feet and stamp tight on the ground all about it. It is tender work, like putting a little child to bed. We next place a blanket of manure over it, and then a counterpane of more turved grass. Surely the little tree can come to no harm, with such generous care. As we leave it we know that we have done for it all that is possible. And the mulberry tree stands proud and erect in its new home.

The walnut is more difficult to

move, for it is much bigger. We will transplant it in the American fashion, by digging a very large pit all round the tree and removing it, roots and surrounding earth, in a large piece of sacking. Thus the tree will feel no disturbance in the nerves of its roots and fibres. I remember watching this being done on Fifth Avenue in New York. Little trees were brought in lorries, the earth surrounding their roots bound up tight with sacking, like the casing of a mummy. There had been pathos in the sight of these little trees, each so soon to be imprisoned in a minute island of mould on the cement sidewalk. They had looked very lonely, cut off so completely from the earth that had fed them while they were yet in the open land. When their roots grow, they will encounter not boundless earth and roots of fellow trees, but resisting cement and the heat of subterranean water-pipes.

But let us see to our walnut. It is hard work, for the roots are deep and wide, and the earth full of stiff chunks of chalk. But at last we have isolated the tree, and Noel and I hold it slightly up while the gardener slips the sacking below it and gathers it up into a bunch above. With a heave the walnut is in the wheelbarrow, and we move it in a triumphal procession to the front orchard, where we place it in its enormous hole and slip away the sacking from under it. We feel certain that it can have felt nothing of all this, for we have never even seen any of the fibres of the roots. But we shall watch very fearfully for its buds next summer.

The extreme damp continues. In the rain we dig and manure the rose bed, preparing it for the newcomers. It is cold moisture, too, and towards evening our fingers are stung with it, as though we had been touching cold metal. Bledlow Ridge is invisible in the mist. The wet wind goes through us. "It's an idle wind," Annie says, trying to comfort us. "An idle wind?" I question her. "But it is far from idle." Annie explains herself. "We call it an idle wind when it's like what it is today. It goes through you without taking the trouble to go round you instead. That's what we call it."

But the starlings do not seem to mind this weather. Perhaps the wet has liberated insects and worms for them. In our field across the lane there is a large flock of them. I wonder whether they have a leader, or if there is some unity of mind and intention in them, that the entire flock should swerve and swoop and circle and rise, its shape solid and unbroken. It is as though they were an enormous black Catherine wheel, deepening in shade as they turn the bulk of their bodies towards me, thinning and lightening as they turn away. At one moment the solid circle falls into the long grass, and is hidden as it pecks at worms; then the flock rises in a rush, the back ones first, with the front ones falling into place in the pattern. This set dance is repeated all over the field. At last there is a great noise, and the whole flock flies towards our garden and alights on trees. One small hawthorn is bowed down by their black weight as they devour the haws.

Towards the end of the month the carrier arrives with a long bundle bound up in straw, that looks like an imprisoned giant. It is the new trees. Within these wrappings, inert and slender, lie elms and chestnuts. It is strange to imagine to what size and weight they will one day grow, for they are now so slight that the carrier can lift a dozen of them unaided. As we look at the young chestnuts we see in them future shade for our treeless lane. The two new elms will overshadow this house long after we are both dead. There can be nothing casual about planting trees. After today the place will never again look the same. We shall have changed the shape of the landscape. As we open up the bundle and lift them out, the trees seem to shake themselves loose and free. They are beautifully shaped, and we turn and twist them in order that we may plant them so that their most interesting side shall show to the best advantage. The

gardener would plant them with their fullness towards the north-east; for, he says, a tree always bushes towards the south and west and falls towards the east. He cuts at the main roots of the trees with a downward slant, that better fibrous roots may form. As we bear them off, a little breeze shakes through the frail branches.

"How they sigh directly we put 'em upright, though while they are lying down they don't sigh at all," said Marty South in *The Woodlanders*. "It seems to me," the girl continued, "as if they sigh because they are very sorry to begin life in earnest – just as we be."

Tenderly we spread out the roots, keeping them flat and separate upon the floor of the hole. I hold the trees upright as Noel throws in the earth.

By evening we have added twelve chestnuts and two elms to the countryside, and have secured for ourselves a stake in the future. And on the branches of the little elms sit surprised sparrows and a singing robin.

DECEMBER

Somehow, Noel and I are always a bit late with everything in our garden. Perhaps we talk with each other too much about our plans, instead of doing the work. All wise gardeners have by now planted their new rose bushes and stocked their beds; above all, they put in their bulbs fully two months ago. But in this last respect we have a genuine excuse. It is the flowery language of Dutch bulb catalogues that I blame for our lateness in planting our hyacinths and daffodils: we have spent so much time in reading them that days slipped by and the bulbs were not ordered.

It is a terrible weakness to take as much pleasure as we do in bulb and tree and seed catalogues. Bulb catalogues, especially, are wonderful "dope". Does one think that the world is wicked, or foolish, or falling to bits, then half an hour with a bulb catalogue will cure one of all this nonsense. For it breathes a spirit of trusting hope. Are we not told that "England has succeeded in restoring her financial balance. Signs everywhere of this awakening to a New Life, greater confidence in the Future"? And thus, "inspired by that confidence we are taking steps not to be found wanting next Spring, when everything will be breathing New Life, and Nature, clad in her new and bright coloured dress, will be calling to us to rise and follow her." So we must sprinkle bulbs all over our gardens, planting them where they fall. We are fully persuaded that it is a good, sweet, kind world, worthy of these thousands of narcissi and crocuses and tulips upon which we are advised to spend our money. The trouble is that when we have finished reading these catalogues, we are firmly convinced that we have ordered our bulbs. It is no wonder that we are late in planting them.

It is a greater act of faith to plant a bulb than to plant a tree. For no matter how small the young tree is, there, visibly, is the eventual form in microcosm. But with a bulb it is different. It needs a great fling of imagination to see in these wizened, colourless shapes the subtle curves of the iris reticulata or the tight locks of the hyacinth. At the time of bulb planting one most nearly approaches the state of mind of the mystic.

This year we are being snobs. Not content with the usual rush of daffodils and tulips and snowdrops, we are planning to have esoteric plants and bulbs in the newly made flower bed. I look through all the bulb catalogues I can procure, to find special tulips from Turkestan, and obscure, unfamiliar crocuses. It is an expensive snobbishness, for there are lovely diminutive daffodils I have never yet seen, and many irresistible Mediterranean crocuses. A strange shaped bulb arrives which is supposed to bloom into an arum from the Himalayas. Annie, who can never stay indoors when anything of interest is going on in the garden, joins us as we unpack these bulbs. She is bewildered by them. "Comical looking things," she sniffs.

"No roots coming out of the bulbs, nor nothing. You say that one comes from foreign parts? Well, I must say, it looks it." So I suppose even the shape of foreign bulbs is alien to us islanders.

We need to throw our minds forward to the time of the flowering of bulbs, for the December days are dull and light-less. For a week or two on end we see no sun, and the days never "let up". Fog hangs over the land, turning the garden into a place of mystery. It is a beautiful thing, clean and soothing. The leafless trees rise out of it, exposing isolated tops of branches in a white sea of mist. The end of the garden is invisible, and where sight is denied to us we live in a quickened world of muffled sound. Through the heavy mist come to us intimations of puzzled cows and sheep and noisy twitters of agitated birds. Drops of moisture hang from each plant and leaf and twig. As Annie goes down the garden to dig up celery, dark footprints follow her, where her boots have trodden into the ground the white beads of wet on the grass.

There is a friendly loneliness about the countryside. Gone from our garden are the summer visitors and the fine weather people. These quiet days of damp are disturbed by no intruders, and the birds hold unquestioned reign in hedges and shrubs. Even we ourselves stray less in our garden. It is remote and the grass is cold and wet to lie upon, and the earth tugs at our feet and pulls us into itself. Everywhere is mud, and it is too sticky for digging. But the memory of the summer's drought is burnt so deeply into us that we laugh with joy as we bring our muddy boots into the kitchen. Even Annie does not mind the "modge", as she calls it. It is a restful, reflective time of year.

Fine weather people miss much in the country. There are subtleties of colour and beauties of form in the autumn and winter that are lacking in the more obvious sunshine of the summer. The blatant greens of August are uninteresting compared with the light on the wet trunk of a bare tree, or the purple bloom on the moist fallen leaves of the chestnut. We sweep the garden clear of leaves, harvesting them for future stores of leaf mould. In the slight wind they patter along the paths like children running home from school.

This is a time, too, of adventure and speculation. We turn in the sods of grass and weed, to decay into fertile food for the plants next year. There is as much excitement about autumn digging, with mist and damp and dark clouds all about us, as there is in the bloomings and gatherings of summer. It is good to feel that these wetting rains and mouldering mists are converting this year's bloomed and seeded grass and weeds into food for next year's beans and peas and lilies.

Neither do the birds share this fine weather attitude towards life. The warm damp of this December deceives them; they sing at dawn as though it were the spring, and thrushes are in full song. Tits crowd round the hopper and the sunflower seeds, greedily eating all that we give them. One envies them their power of gymnastics as they balance upside down at their meals. It must feel good to be able to move about in so many directions and dimensions. We have noticed a greater number of them, and we wonder if pairing is already starting; they are restless, too, as though the year had turned and they were feeling the approach of spring. But the loveliest creature of all in our garden is the green woodpecker. Throughout the year he comes and goes, obeying some rhythm of food supply, of which we know nothing. Just now we hear his chuckle each morning at exactly eight o'clock, and hurry to the window to look at him. I wonder what Nature was doing when she gave him his exotic colouring. Why should he have been clothed in such bright beauty for this dull light? His rightful place should be the hot sunlight of tropical trees, a fitting comrade to the parrots. Here he hops about in the tangled grass of the orchard, busily pecking at something. We hope it is ants,

for he is especially attached to them, and laps them up with his four-inch long tongue. Now he climbs the poles round which we are training clematis, ripping off the old bark in his search for insects. He is the most welcome inmate of our garden.

The worms disobey the calendar. At this time of year they should be about three feet below the surface of the earth, but so warm is this December that not only do they dot the lawns with their coiled castings, but they mate. The coupling of earth worms is a beautiful thing to watch. The pair emerge from the ground at points about nine inches apart and move toward each other, until they embrace, leaving half their length still under the earth. Thus they remain, blissful and unmoving, until the slightest vibration reaches them, when they withdraw from each other into their holes; for earth worms are shy lovers. The grass is patterned with them.

Now we change our gardener again, and Lacey comes to us. He is rhythmic, and his voice is earthy and rumbling. He digs with long slow movements, in harmony with the earth he touches. His breath in the damp air is like white smoke. We no longer feel that there is a discordant atmosphere among our plants and trees, and we are happier. Even Annie likes him. "He's rough, he is. That's what I like. You can always get on with them when they're rough." She puts her feelings in a somewhat crude way, but they are probably the same as ours.

The life of the garden is very gradual. There are none of the dramatic divisions of the seasons that we are told to expect. As the month runs through, we gather a strange assortment of flowers belonging to summer, autumn, and spring, all at a time. For the roses are still in bloom, with their muted autumn colourings: white is turned to pale green, yellow into dull gold, pink into flushed fawn. Auriculas show a clump of purple inside the cup of their leaves. Wallflowers and marigolds rival each other in brilliance, and primulas and chrysanthemums blossom in the same bed. Only the mauve of fungus, half hidden beneath drifts of decaying leaves, reminds us that it is still autumn.

But this year the merging of the seasons is exceptional, and frightens gardeners. For all around are the first spears of spring bulbs. Scylla show above the earth, and tulips and daffodils point upwards. And, fearless of coming frosts, the iris reticulata is almost in bud. Villagers shake their heads at the "unseasonable weather", and I recollect the terrors of my childhood, whenever the seasons behaved in an unordered way. My old nurse was always reminding me of the Day of Judgment, thinking thereby to make me into a good girl. "And you'll know when it's coming by the seasons," she would thunder at me. "For they say that at the end of the world there will be no difference between winter and summer, spring and autumn, except for the falling of the leaves." I would tremble as I watched the weather, and gained quite an unjust reputation with my parents, who were ignorant of all this, for being an observant lover of nature. I laugh to myself as I wonder if anything of this old fear remains with me as I watch the spring bulbs thrusting themselves up in December? I imagine I merely see visions of iron frosts in the early months of the year.

There is beauty in these signs of growth, proving to us that there is no stopping point in the garden, no end to the year. Two months ago, in October, the branches of the hazel were heavy with nuts. They break now into tight buds of catkins. The push of growth is strong in the cherry trees, where sturdy knots of buds clump ready for next year's blossoming; lilacs swell.

It is hard to realise that Christmas is upon us. The clouds tighten into shapes, and lift, and in the sudden darts of sunshine a bee hums on a marigold. Creatures of habit, we need the traditional frosts and snow, and it is only by a few signs that we are certain it really is Christmas-time. Over the fence at the bottom of the garden our neighbour's geese have suddenly disappeared. For weeks now I have enjoyed the sight of them, feeling with them the cold and the wind, and amused at their way of roosting at dusk, side by side, like two cold white china ornaments on a mantelpiece. Now I miss this whiteness. At night time carol-singers come to our garden, dimly perceived in the blackness, whereas, by rights of tradition, they should stand out dark against snow. The milkman gives Annie a bunch of holly. I go into the garden and stare at the fir trees; it is their time of growth, and they decorate their branch tips with bursts of new pale-green needles. Somewhere at the back of my mind, vague and half formed, is an instinct further to decorate them. The old urge for ritual enters into me, and is unsatisfied.

In this sunshine of Christmas-time we can work in the garden. The untoward warmth has started the grass growing. Among the worm castings on the front lawn, it is choked with moss. Shortsightedly we have let it spread, unhindered, and now it will give us infinite labour before we can destroy it all. We spend hour after hour with the spring rake, scratching up the moss. At first it fascinates us to watch the moss come up with the teasing of the rake, and to amass it by wheelbarrow loads, but after some time the novelty wears off.

It seems endless work. As we scratch and scratch, more and more moss comes off, till the remaining lawn consists of a few brown blades of grass. We understand now why we have delayed clearing it; we have known what the moss-free lawn would look like. But this unsightliness will not last for long, if this warm weather continues to force the new growth of the grass.

And then we betray our garden. We leave it for a time, as hard-hearted as the fine weather friends at whom we scoff. I find that one of two things may happen on my return to the garden after an absence. If I am lucky, my senses will be keener from the separation. Subtleties that I have not before noticed will leap out at me; the gradual growth of the plants will appear sudden and dramatic, and I shall be super-sensitive to colour and light and form. But I have sometimes come back to the garden with my mind so completely out of focus, that I am incapable of seeing or enjoying the sky and the smell of the earth and the shapes of growing things. This is a wretched state, and I am deep in it now. I watch the sky after the rain, and see gold edges to the ragged grey clouds. Quivering light floods the wet garden. Yet I am unmoved. What is worse, I am aware that I am unmoved, and that I am wasting all this beauty. I walk round the garden, hoping to grow in sympathy again. I feel beauty but as a reproach.

The one thing I must do is dig.

There is great healing power in digging. This is so much the case that one is tempted to wonder if any actual electrical power comes up to one from the earth. Perhaps the benefit is merely from the rhythmic movements of the body. At any rate, however sulking and rebellious one may be at the start, sensitiveness creeps up the fork into hands and body and legs. Finally the brain surrenders and one is again at peace with the garden.

There is much to see just now. Winter jasmine pours over the house like a rushing yellow river. Christmas roses gleam wan and ghost-like under their glass cloches. Ivy berries clump dull and black in the hedges. I am reminded of a man we know who destroyed the whole

of his hedge, because ivy climbed over it. He explained that there was something obscene about the ivy plant, with its snake-like roots and stems, and that it disturbed him. Some puritan complex was at work. We argued with him, and reproached him.

"And you feel nothing at having destroyed the flowers and berries?" we asked. "Have you forgotten how the bees crowded on your hedge for the honey in the summer, and how dark the berries glowed in the winter?"

But his fighting spirit was up. He squared his shoulders and confronted us.

"There aren't any flowers on ivy," he declared. "Never. None." And we failed to convince him of his sin. Let us at least hope he is more at peace, now that the obscenity no longer confronts him. But this is poor consolation to us for the loss of his hedge.

Now, too, is our chance to learn the shapes of the new trees before foliage obscures their structure. The bones of the chestnuts are straight and thin, like the shafts of an antelope's leg; but there is no uniformity of shape among them, and we get to know the height of this tree's trunk or the way in which that tree branches out. The new green cooking apples are tall and thin, and bunch at the top in a nest of twigs. The elms are feathery and less simple of structure. We shall enjoy them better next summer from knowing their disrobed form.

Lacey prunes the fruit trees. Tenderly he cuts back all the branches except the leaders, leaving on these always four buds. It is gentle, reproving work. He has had his secateurs for over twenty years, and guards them as though they were sacred. This respect for the tools of his craft is usual in the countryman. We go the round of the fruit trees with him, secretly jealous that he has taken the pruning into his own hands; for we had meant to do it ourselves, and have only missed our moment by being away from the garden lately. This is the second time we have been baulked this year, for we had failed to prune the roses. There is chancre to be cut from some apple trees. The drought has withered branches of a greengage. A poplar has burst its skin all down one side.

"It's a good thing it done that," says Lacey, "otherwise it would have strangled itself, as trees sometimes do."

The year closes in rain and lack of light. We are house-bound, or plunge about in grey mud. It is almost impossible to avoid feeling depressed, and we are tempted to rebel. But we grow reconciled to this bored, bound-in feeling, as we realise that escape from it would deaden the contrasts of the year. May stands out brighter and more desirable against a background of December. It is possible that we need a period of gloom in the same way that the darkness below the earth is necessary for the growth of seed and bulb. And sure enough, as we walk round the garden on the last day of the year, pulling our feet with difficulty from the slippery mud, we find a clump of primulas in flower. The year tastes best to those who do not fear the bite of the wind or the drench of the rain.

JANUARY

THE FIRST THING I HEARD on New Year's Day was the spring "sawing" song of a blue tit. Was it ever before so early? It was a fitting start to the year, and my mind leapt with this "sawing" sound to the sunshine of April, telescoping the intervening months of frost and darkness, and cold rain. This is the pleasant side to the precociousness of everything this year. But in this open warmth the sap flows, and the trees are not getting their required sleep. There is a feeling of unrest in the earth and things everywhere are stirring. Buds appear on the loganberries and the black-currants. The garden gives precious hostages to the frost.

But growth is very soon arrested by the winds. I doubt if any garden suffers more from the winds than ours. Here, on the slopes of the Chiltern Hills, we offer ourselves to the gales with an air of bravado. Our hedges are a sheepfold, protecting the little creatures inside them. While we must layer the hedges to make them thick, we must keep them as high as possible to lessen the force of the winds that sweep cruelly over them into the garden. If one could isolate oneself from a feeling of responsibility towards plants and trees and birds, wind excites and stimulates. In the deadness of sultry, windless summer days, even the thought of it braces one. There is nothing more satisfying than to lie in bed at night, secure and warm, with a whistling wind outside. Windows creak and flap and grumble; one's senses are limited by the darkness to hearing, and so the moan of the wind lifts and falls, strengthens and diminishes with a range of sound that is unimagined during the daytime, when hearing is distracted and tempered by sight, and wind means the racing of clouds across the sky. In this keen hearing of night-time we can distinguish the roar of the wind against the elms in the lane from the swish of the swaying poplars. But, to enjoy wind one should not own a garden. One feels remorse at this exhilaration, when the northeast gale dashes the small plants low to the ground, where they lie flat and broken. It hardens the surface of the earth. It bends and snaps young trees. Fearfully we watch our newly planted trees and strengthen the supporting stakes. The strips of cloth that bind them to the stakes untie and flap like pennants. At the feet of the trees are large gaps in the earth, where they have struggled and bent and worked loose in their holes; we fill in these holes and stamp heavily on them, securing the trees for a few more hours of battle. A young acacia snaps. We have tied it too tightly to its stake, and it was unable to save itself by swaying with the wind. Three years' growth is destroyed, and the tree stump stands reproachful and jagged. Branches of a laburnum break. Battered fragments of plants are blown across the garden.

Even the glass cloches are torn from their positions.

But it is the birds about whom we are most concerned. The gales are so forceful that even large heavy creatures like the blackbird are blown out of their course. We fear lest they should be dashed against tree trunk or telegraph post. To our distress we see a sparrow bang against the dining-room window, falling to the ground in a dazed state; but he recovers and flies off. In the hawthorn trees along the bottom hedge thrushes sway against the sky like stars seen from the deck of a ship at night. The tits' hopper outside the window swings like a roundabout; and on it sit a great tit and a blue tit, their feathers ruffled outwards by the wind.

It is interesting to notice the different manners in which trees wave in the wind. The ash is steady up the trunk to its shoulders, and then shakes in a circular movement. The elm is less original and waves all over its body, each branch managing its own motion independent of the tree as a whole. Poplars bend over from their slimmer trunks. Our new chestnuts lean from their base in the earth, but their trunks remain rigid.

We garden in the wind. The rain lashes our faces like spikes. The north wind goes through our bodies. Though we pile on layers of clothing, it is impossible to keep warm. But the new hepaticas have arrived and must be planted. Hurriedly we put them into the earth, that their roots may be protected from the winds, and put glass cloches over them. These cloches give such a professional, intentional air to a flower bed! They make us feel that our garden begins to look like other people's.

The Christmas roses, with their appealing white, are safe under them. The purplish green ones I brought from Aunt Sarah's garden still refuse to flower, though they have had three winters in which to consolidate after their upheaval. I believe they are more shaken by removal than most plants.

As these winds rage, and we shun the garden, so do we enjoy it from the house. One learns a great deal by standing at upper windows and looking down upon the garden and the birds. The blue on the top of the tits' heads is extra bright, and one can see things that would otherwise pass unnoticed. I think we are too much inclined to look at things from our ordinary standing or sitting eye level. I remember shocking a dignified elderly matron when I suggested that she would see and enjoy the autumn crocuses much better if she would lie down on the grass, as I was doing, and look at them all at once, against the sun; the distances between the flowers were less and the mauve shone out brighter. In the same way, the colours of a landscape are more intense, and the component parts of the various colours much more evident, if one looks at it upside down, between one's legs. I have often wondered why this is, and have never had it explained to me; but it is a fact. So we look down upon the garden and the tops of our young trees, and decide how wet or how dry our earth is, and plan for our summer's flower beds. Even greater fun is to stand at an upper window with strong field-glasses. If one has a sense of decency towards the birds, this is out of the question,

but if curiosity outweighs this sense of decency, as it does with me, then one can watch all the little intimate family matters of the birds' lives through the year; I have seen mating, feeding, quarrelling, and sitting. I have attended a particularly shy thrush in her three weeks' vigil on her nest, though we shunned the back path lest we should disturb her. I have learnt how the swallows on the wires twist their heads round as they search their bodies for insects. I have seen the chaffinch clearing a plum tree of green fly. And if I have by chance discovered a pair of blackbirds in the delicate details of courtship, I have never hindered them.

We enjoy talking with Lacey. We merely mentioned to the last gardener what we wanted done and left him, but now we make any excuse to go round the garden with Lacey and listen to his rumbling, earthy voice and watch him work. We find him planting an avenue of poplars, taking advantage of the sun which has thawed out the earth. Noel eagerly helps him, forgetting the work he should be doing indoors. As the trees are put in, the trench is filled again from the mounds of earth that have been heaped on the grass for weeks past, while we waited for the delivery of the trees. Now I see that my fears were justified. I had warned Noel that these mounds of earth covered our snowdrops, but he had assured me that, fortunately, I was wrong. Today we see them, a pitiful sight. Here are scores of emaciated, light-starved flowers, their leaves pale yellow for lack of chlorophyll, their stalks lanky from their valiant push upwards towards the surface of the earth, which was heaped three feet on top of them. We feel that we have been cruel, and we are ashamed to look at them, or to let Lacey see them. But Lacey's attention is discreetly occupied with the poplars. Perhaps it is subtle courtesy towards us. Anyhow, he stamps down the earth round the roots, with his big, heavy limbs, scraping the sticking clods from his boots with his wooden "lazy boy".

We go the round of the vegetable garden with him, for it is high time we ordered seeds. We plan where the parsnips and carrots shall go, and how many rows of peas and broad beans we shall need. He will not allow us to grow our own green-stuffs, but insists on starting them in his allotment and giving us the young plants. Our soil swarms with spring tails and they would devour the young cabbages. I ask him what a spring tail is like, for I have never heard of them. He pounces on one with his enormous hand in an empty vegetable bed, and shows it to me in triumph, a tiny insect that hops. He teaches me about "big bud" on the black-currants, as he removes it from the bushes. I shall soon get to know quite a great deal about a garden.

The flower seeds are ordered, too. But it is hard to limit one's choice to what is really necessary. There is nothing in which one feels as much greed as in the choosing of the seeds of annuals. We want them all, even if we only have a small quantity of each one.

I make out my list, and Noel his. We must have the old favourites of last year, and we must try new ones. When they come, we shall probably not have the space in the garden for a quarter of them. In one of Noel's drawers, tucked away behind neckties and collar studs, lie various packets of unused flower seeds, purchased over a period of three or four years. Several packets are destined to swell that number this year. We wonder if the seeds have lost their potency after this long time. We mean one day to scatter them in our field across the lane, to see what will happen. And in this same drawer, in an old envelope, are some seeds that we brought home from Majorca five years ago. They are of an especially lovely dandelion that we found growing there on the mountain sides. It was a very pale lemon yellow, backed with delicate stripes of brown. When we first made this garden, we had a plan of scattering these Majorcan dandelions all over the orchard, and bravely we tried to make them grow. Two plants came up and flowered during a summer, but the English winter destroyed them, and we have now lost heart. The remaining seeds wither and grow old. Other seeds from Majorca are added to our list of failures.

There is one advantage about winter in the garden. We get the utmost enjoyment out of the few rare flowers that blossom. During the rush of late spring and early summer, when flowers tumble over each other, there is not time to watch and analyse the beauty of each plant. Now we look at the sprouting spear of the iris stylosa and watch it intently through its stages. The spear begins to fatten and show a tinge of violet in its substance, the bud is straight and full and graved with purple lines round its tightly bound petals. We cut one and bring it indoors. The top part thickens; and suddenly, even while we are looking at it, a petal uncurls from the tight roll and leans back slightly, to show the violet glow of its inner colour. Back and back it leans, and the rest of its petals expand, till it is a parti-coloured pattern of violet and ivory. At last, after a few more hours, the entire flower is free, all limp and expanded and relaxed. There are yellow splashes and white lines upon the deep, rich lilac-violet of its petals; sheath and stem show a wonderful clear, transparent green. Surely this flower, with its perfume of green grapes, is one of the loveliest of the whole year. Why does it perform the caprice of blooming in the dead of winter? But we are glad, for else we should never have remarked all its beauty.

Hard frosts come. The ground is unworkable. We put cloches over any delicate plants that we cherish. The new bird table is erected near one of the windows, in order that we may watch it. And it does not take the birds long to get accustomed to their new food supply. Each morning there is a scramble as Annie takes out water and bread and fat. Robins arrive, and then thrushes. Some of the tits

desert their hopper, hoping for a change of diet. They are bold little creatures, and they chase off the robins. But it is the sparrows who dominate the bird table. I hear Annie shooing them away:

"Go on, you 'orrid lit'le things! Here, you, robin, hurry up. Now's your chance . . . Quick, I say to you, quick!"

It is strange how snobbish one is about birds. Even Noel surprises me; he urges Annie not to frighten away the sparrows, lest their obvious fear of us should keep away the "nobler birds". We, who pretend to love all birds alike, seem to treat the drab little sparrow as an outcast. Is it because we feel that he will always be able to make a way for himself without our help? I am afraid it is lack of interest in him. A little old lady in our village is a much finer character than we are. She always takes out with her on her walks a large reticule full of pieces of bread. She is to be seen sweeping down the lane with her arms flung out as she moves, strewing the path with bread, as though she were sowing seed. "I must feed my birds," she explains. And the sparrows of the entire neighbourhood fly after her. It is a lesson to us, we who have a proprietorial feeling towards the birds that are within our own four hedges.

As the frosts loosen, we work again in the garden. I weed the flower beds and Noel collects the wood ash from our fires, where we have let it accumulate since early autumn. He distributes it over the perennial bed where the bulbs were planted and on the bulb square in the middle of the front lawn. He looks happy, and says he enjoys doing it, for it makes him feel benevolent. The bulky leaves of the autumn crocuses begin now to show. We search for early spring things, knowing that in London, and even in sheltered gardens in the plain, snowdrops are out in thick dripping clumps, and crocuses are inches high. And towards the end of January we discover the purple gleam of a crocus, and find a few young snowdrops among the flowering aconites.

We are apt to get very jealous at the way other people's gardens always bloom so much earlier and more abundantly. In the same way, we resent it when we find that our garden lacks so many flowers that other people have. But this is stupid of us. For it is in the gradual building up of a new garden that most pleasure is to be found. However pleasant it may be to step into an old established garden and watch the flowering of mature plants, nothing can equal the stimulation of the first years of struggle. I know a woman who had a marvellous herbaceous border. It looked exactly like the outside cover of a seed catalogue. I was filled with envy.

"How long have you had this garden?" I asked. "It must have been years and years and years."

She boasted to me that this was her first summer. I was amazed and envious of her skill. But I learnt later from some neighbours that she had put the entire place into the hands of a landscape gardener who had imported earth, plants and all. Suddenly I felt sorry for her; she had known so little of the joys of gardening, for she had not had to fight. She had never raised seeds and watched the first green shoots thrust themselves up through the earth. Slugs had not bitten off her seedlings. She had never waged war against the wireworm. Her planning had never excited and troubled her. For everything had been done for her. So, when I find that we lack some flower that we should have, I remember that we have ourselves watched this garden grow slowly, year by year, from rude meadowland to flowering beds and bushes. And I decide that it is fun to notice the omissions each season, and to determine how we shall enrich the garden next year.

We should never take our gardens too seriously. It is hard to curb ourselves in this, if we have any love for our plants, even as it is difficult to take a walk round the garden without pulling up weeds. But too professional an attitude is apt to give us the same taut, strained feeling that comes into the faces and lives of all specialists. It is better to have a few weeds and untidy edges to our flower beds, and to enjoy our garden, than to allow ourselves to be dominated by it. To be able occasionally to shut our eyes to weeds is a great art. Let us relax in our gardens, and as a dear old countrywoman used to say, let us "poddle" in them. We waste else the very beauty for which we have worked.

FEBRUARY

A S I WALK ALONG THE ICKNIELD WAY I meet an old countryman.
We stop and talk about the weather. Candlemas has been fine
and sunny: this, according to rustic weather lore, is a bad sign, and
portends a hard winter before us. My old man shakes his head at the
budding bushes.

"It be so cold, mornings," he says, "I wouldn't be surprised if we
had a picture."

In this temperate climate snow is so gentle and friendly that
country people about here call it "a picture". Presumably it brings
before their eyes visions of Christmas cards with snow scenes and
red-breasted robins. Towards it there is no fear.

My old man is right. By midday the snow has started, blown
horizontal across the land by the violent winds; it lies in a thin scatter
on the hill-tops, transforming them in our imagination into high
mountains. By nightfall it has begun to settle, and Noel comes in
from the post with snow on his shoulders and boots. I listen in bed
to that absolute silence without, which comes upon the earth only

when it is covered with snow. I know I shall awake next morning to a whitened world.

What excitement we feel on looking out on to the garden in the snow. It is one of the only sensations of our childhood that is not blunted by maturity. Still we want to leave our mark on any smooth expanse of snow, to ruffle it, to jump about on it. However sedate we may grow, we never emerge from the childish longing to write our names on the whitened lawn with a stick, as though it were sand by the seashore. It is a pity that we have so little snow.

The wind plays strange pranks with the snow. It changes the contours of the garden, giving us banks where we had flatness, lying in drifts which level the slopes. It raises a mound over a stone and drapes the water butt. The little juniper bushes stand wrapped in white scarves that trail down to the earth on the windward side. All about the garden are strange white lumps, blown into being by the wind: they look like sheets thrown over bulky furniture at a spring cleaning. As the sun comes out, it enforces these weird shapes of snow with pronounced shadows. The snow turns the white of the Christmas rose under its protecting cloche to pale green; it heightens and burnishes the greys and browns of sparrows and thrush. Over the fence our neighbour's chickens glow; the gold and crimson and green of their cock sing against the white ground. For dull colours that would pass unnoticed against the usual background of earth and field become infused with life when they stand against snow.

I had often wondered at the terror that is felt at snow. There is nothing frightening about it here, in the South of England. It is a delicate, lovely covering that beautifies everything it clothes. It keeps the little plants beneath it sheltered from frost and wind. It melts as moisture for the earth in spring. It turns the world into a playground for child and dog, and muffles the ugly sounds of civilisation. But, after spending a winter in the Canadian woods, I understood. For

one day I had stepped off the trodden path through the bush, to flounder to my shoulders in snow. I had lost my way in a snowstorm, three minutes from the hut in which I was staying, when the falling snow was so thick that it hid from me ground and sky and tree, and obliterated the guiding lace pattern that, spider-like, my snowshoes had spun behind me as I walked. In such a country snow terrifies.

But in two days our gentle snow has disappeared, and already there is a smell of spring in the air. The temperateness in the weather here may be comfortable, but it lacks drama. The coming of spring must mean more in a country which is locked in ice and frost for months on end, where the ice booms on the lakes and rivers, and there is suspense until it starts to break. I remember, too, the drama of the melting snow in the Tyrol, where, after a winter in which the whole land was white about us, the sunward surfaces of the mountain sides darkened to earth in a day, exposing a world of little budding plants, all warm and thriving, and prepared to flower within a week or two. The melted snow poured from roofs and gutters, splashing upon us as we passed through the village streets, and rushing down the meadows in little roaring torrents. The seasons stepped forward, straight from winter into flowering spring, with none of the little backward turns and loiterings that accompany the march of our English year.

Yet to us, who are accustomed to nothing more violent, spring seems to come suddenly. We have weeks of frost, when the earth resists all effort to move it, and ice lies thick on the water butts in the midday sun, and birds gather at the kitchen window for food. But one day we turn to each other and say that the spring has come. We can almost imagine that we know the exact half hour of its arrival, as though it were a visible guest. I do not know through which of our senses we are first conscious of it. I think it may be through our sense of smell. For spring has an intangible but unmistakable smell, even though we are at a loss to state of what it is composed. Each season announces its arrival with its own individual smell. Who can doubt of the first moment of autumn, with its knife-like tang of wet leaves?

The birds are aware of the coming of spring, even as we are. Starlings chatter, crows, blackbirds and thrushes seem to be everywhere about us. They are all very restless. Nesting has not started, but there is a suggestion of building in the movements of birds as they survey the hedges. Courtship broods over the garden, and with the instability of spring-heightened emotion, birds fight and quarrel. A blackbird chases his hen across the orchard; he is a few yards behind her, as she dodges among the apple trees. He overtakes her as they fly towards the woods, and are lost to view. The hawthorn hedge shelters another blackbird pair in the heat of a violent love quarrel. In the sky, above our meadow, larks sing. I feel a sudden rush of pride in the ownership of this untilled land; it is not that I value possession, but I am proud to be landlord to the lark who pays for his home with such boundless song.

At the bottom of the orchard sits a clump of coltsfoot. As the dove with the olive leaf told of the end of the flood, so does this pale gold flower signal the close of the winter. Bravely it blossoms before its protecting leaves appear, an obscure little figure of low repute. Yet it has exquisite beauty of form. We rush after cultivated flowers and overlook the grace of the tiny wild ones. I remember a purple carpet of wild crocuses on the rocky coast of Corsica; we did not think less of them because they were small, for each flower was only an inch high. Scattered among them, and towering a few inches above them, were minute wild arums; they were as interesting in shape as the greenhouse variety. Noel especially fights this ambitious craving after size; it is the shadow that the annual flower show throws before it. Each year he tries to stop the gardener's drastic pruning of the roses:

"I will not have these enormous vulgar blooms. What I want is more roses, and smaller ones. Next year – "

But next year the gardener again gets his way. Lacey tells us of his flower show successes.

"And I had one dahlia eight inches across, and runner beans eighteen inches long, and – "

But Noel has fled. He finds refuge among the wild rose bushes in the lower hedge.

This little coltsfoot has set me thinking about the pull back to the wild that goes on ceaselessly in our garden. I do not know whether this is more evident with us than with most people. I think perhaps it is, and that it is partly because of our tolerance towards flowering weeds among the orchard grasses. A great deal must be due to the youth of our garden. Wild meadowland cannot be entirely changed into cultivated garden in four years. And why should we mind? If we can keep the pink and white bindweed from strangling our roses, should we deny it a place among the uncut grasses? If our hedges contain wild rose and bryony, should we forbid the cuckoo-pint its home in the dark of the hedge-bottom? A garden should never forget its origin. It should never become so remote from field and lane and coppice that the link is smashed. Part of it should be allowed, and even encouraged, to lean back to the wild. There must never be a cultural fence between one's garden and the neighbouring meadows.

The seeds have arrived. Lacey calls me into the tool shed with a glow in his eyes. Proudly he has arranged these small packets of seeds along the top shelf, supported against the wall; snapdragons and asters, dahlias and stocks, sunflowers and salpiglossis stand among peas and beans, carrots and leeks. I feel the usual sentimental wonder at these little packets of dull-coloured seeds. We sow snapdragons and asters in boxes of sifted earth and place them in the shelter of the frame, which we have at last kept inviolate from the cucumber. Lacey plants the broad beans. Lettuces and sweet peas already show green under the glass. The empty vegetable plots stand dug and tidy, prepared for the first sowing weather of next month.

In this earliest spell of spring we rush to do the weeding that was impossible under the iron grip of the frost. In spite of the optimism of the little coltsfoot, winter may recur at any moment. Even in this spring sunshine I feel the earth grow harder towards nightfall, until the roots of the weeds are imprisoned in the ground, and the stems snap and I am defeated. We spread bone-meal over the flower beds. Lacey sprays the fruit trees against blight, breaking off from time to time to look for his pet "spring tails" in the vegetable beds. In a corner of our meadow we plant a few handfuls of chestnuts, saved from the autumn. Will they germinate? Shall we one day see a miniature chestnut grove of our own sowing, even as my old Aunt Sarah lived to sit in the shade of cherry, peach, and almond trees that she had grown from fruit stones? We are giving ourselves a long distance view of our garden.

Two energetic bullfinches are busy in the orchard. They have discovered that the swelling buds on the fruit trees are juicy. With enviable concentration they perch on the branches and eat. Wisely we have not yet pruned the gooseberry bushes, and this is very efficiently done for us by the greedy bullfinches.

We have a new neighbour. Over the fence at the foot of the garden a young she-goat has arrived, heavy in kid. Save for the weight she carries, she is an elegant little creature, and we hasten to make friends with her through the wooden stakes of the fence. Old Nellie, the big hoary matron goat, is jealous, and pushes the forward young thing aside. They are nice neighbours to have, so beautiful of form and gentle of movement; and they make none of the insistent demands upon our time and energy that human neighbours would do. Today, in the bright warmth, a hen has hatched out eleven chicks among the goats; they emerge from under her puffed-out feathers, to walk about in the light that is nearly as golden as they are themselves. Our neighbours, too, appear to enjoy this burst of spring.

Then, one morning, we see a large flock of white birds over in our meadow. Fearfully we realise that they are seagulls that have flown far inland for shelter. This must herald stormy weather. For we are nearly forty miles from where they gather at the Thames, in London, and about eighty miles from the nearest point of the open sea. They are strangely out of character on the green background of meadowland. We wonder what they will eat. Noel suggests that we should acquaint the little old lady of the village of their arrival, that she may feed them with her sparrows. They look dead white as they circle in the mouse-coloured sky. There is a smell of snow in the air. So, it would seem, our burst of spring was not to endure.

Storms come and dash the crocus to the ground. The hedge-bottoms are strewn with the white petals of the blossoming wild plum. Little can be done in the garden, and Lacey trims and layers the meadow hedges. It is good that he should do this early, before the birds have begun to build; he might else slash at a nest. We only hope that no birds have cast eyes yet on this hedge for their building. It is a high, tangled line of bushes, a fitting place for shelter. He is drastic in his cutting before we can stop him, but we know that he is doing right. It is hopeless to be sentimental about the individual bushes that compose a hedge. But I have to put up a strong fight for an elder tree; Lacey says that elder is poisonous to the rest of the hedge, and he would grub it up. We manage to save it from his hook, and it stands high and proud among the levelled thorns and brambles. The little spindle trees, too, are allowed to stand. Layering a hedge is a work of great skill, and Lacey delights in it.

He likes to feel that he knows exactly how to split open a branch and bend it low, with just enough attachment left to the outside bark of the bush that the sap can rise and feed the fractured branch, and leafy shoots spring up from its new bent position. We gather up the lopped branches, blackthorn and dogwood, wild plum and bramble, and pile them high for a bonfire, stumbling as we go in the rabbit holes and ditches that dot the meadow. I am constantly being reproached by a utilitarian-minded Russian friend, for leaving this land uncultivated. "It is a wicked waste," she tells me. "No land should be unused." But, wicked though it may be to leave it untilled, I am certain that we should not feel perfectly happy if we were to deprive our skylarks of their homes. Besides, this rough grassland, with its colours that change and flush towards hay-time, gilded with ragwort, bronzed with seeding sorrel, and its surface that rises and sinks with the wind, is a fitting foreground to our view over Bledlow Ridge. Its very roughness is a corrective to the civilisation of flower beds and lawns.

The hedge trimmings make a splendid bonfire, leaving behind such a deposit of wood-ash that Noel brings across the wheelbarrow, and fills several sacks with the precious plant food. For we must continually feed this hungry earth of ours. The villagers tell an amusing story of the insatiable appetite of our chalk soil. A certain man who was digging in his allotment grew hot, and took off his coat, and went home that evening without remembering to pick it up from the ground where he had laid it. When he went to fetch it next morning it had gone. The hungry earth had eaten it. It eats up the iron in the soil, too, so that we have periodically to replenish it. Otherwise, from time to time, our plants and trees grow anaemic, and the green of the leaves turns yellow.

But in this wild, stormy weather, rivulets of gold crocus run through the garden, and clumps of snowdrops stand like patches of unmelted snow under the trees. The almonds are covered with tight buds of deep pink. Hyacinths push above the ground. The month gives us a pageant of clouds. I had not realised that we were especially rich in the beauty of our skies, until a young American friend came for the first time to England. He said that he would sit at his window in

Hampstead for mornings on end, unable to work for the beauty of our clouds. "They are so large, so rounded, such beautiful shapes," he would tell me. And I think he was right. Our west winds pile up for us mountain upon mountain of cloud that give drama to our skies.

Our plant of stinking hellebore thrives. Thinking to enjoy the beauty of its flowers, I break off branches and bring them indoors, to the warm atmosphere. But, after an hour of heated comfort, they wilt, and I can only revive them by placing the jar outside in the garden. With a few minutes of exposure they stand erect and strong, restored by the cold and the frost. Some flowers actually need the very conditions that destroy others.

As we sit indoors, looking out at the cold rain, nothing seems to move in the garden. But let us go out and look intimately at it, and we see that in spite of this checking weather, growth does not stop. The perennial bed thickens. Scabious and phlox, delphiniums and veronicas sprout above the earth with their new crowns of leaves. Japonica bursts into flower. The stinging scent of the American currant already comes from shaped buds. Spring is upon us, and will not be hindered by winds or rain, or scurries of snow.

MARCH

WE HAVE JUST VISITED SOME FRIENDS in a sheltered garden in Oxfordshire. As we went in at the gate we moved forward fully a month. Here were flowers in full blossoming: anemones, daffodils, scylla, cyclamen, and hepaticas clustered warm and still. I come back to our garden full of apologies to our struggling, wind-blown buds. How can we hope to grow anything in this cold, exposed place? I even wonder if we have any right to subject these creatures to such unfriendly conditions. And then, as I watch the tight buds of hyacinths and hepaticas and the short green spears of daffodils, a change comes over me. Perhaps we are, after all, to be envied for the lateness of our seasons. Is it not like the privilege of hearing over again music that has come to its end? I have had the pleasure of seeing the spring flowers in our friends' garden, and now I can travel back in time and watch the annual unfolding of buds. I am, in fact, enjoying a duplicated experience, and this will go on throughout the year with roses and lilies, strawberries and chrysanthemums.

The winter returns. Black frost tightens the land. It is undramatic as compared with the hoar frost that silvers the garden. Visually one is hardly aware of its existence. Only the ice on top of the water butts show us, from the house, that the weather is not open. But outside in the garden the leaves of plants curl up and shrivel. The little deep purple iris reticulata, with its rich orange streaks, has been bent down by the frost. On the lawns I find several frozen worms. The bird table is a busy flutter of wings; traffic there is constant, for no bird just now can find worm or insect. But spring has already clothed the cock chaffinch in his brightly coloured nuptial coat. His hen looks dowdy against him in her subdued fawn and grey. And, though the land is lightless, the blue and yellow of the tits seems gayer than usual. The green woodpecker, too, is a brilliant splash of colour as he races up the trunk of the elm tree in the lane.

There are some things that do not seem to mind this persistent frost. The shoots of the frail young bulbs stand as stiff and erect as possible, and the almond trees toss their pink blossoms in the wind, lit by their own inner light. The elms are rosy-edged with buds.

The hepaticas are out. This is the first year we have had them, and we have watched impatiently for their flowering. Now they dot the new bed with their bright pink and blue and white. The pink and the blue occupy exactly the same position in the scales of their respective colours. It is exciting to add new members to our family of plants, and this coming year we should see many unaccustomed flowers in the garden. We are planning to add esoteric plants to the esoteric bulbs that we have planted in the new bed. For weeks past we have searched the nurseries of England for unusual plants. Probably this esoteric bed of ours is an outlet for some repressed,

unsatisfied snobbishness. We know perfectly well inside ourselves
that our garden is inferior to most others in the quality of its flowers.
Here on the hills we can never get such large, grand flowers or
vegetables as the people on the plain. Often in bicycling have I raced
at top speed past waving plumes of carrots or enormous clumps of
primulas in low lying cottage gardens, trying to pretend to myself
that I have not seen them; for our carrots would be about half as
large, and our clumps of primulas insignificant by comparison. By
way of compensation we may be allowed the snobbish luxury of our
esoterics. But I wonder very much if any of them will come up. Our
pleasure will probably be only in imagination, and we shall have

to fall back on marigolds. And why not? We are getting so much pleasure already out of the prospect of these plants, that it will have been worth while, even if no single green shoot should appear.

We choose some plants for the beauty of their names. It is an amusing reason, and I remember selecting a holiday itinerary on the same principle. I bicycled through France with a romantic-minded uncle, and we would spread out the map and decide where we were going entirely on the strength of the names of the places. We zigzagged miles out of our way because of the mellifluous sound of a tiny village, only to find when we reached it that it was an exceptionally dull place. Yet we continued thus in our mad holiday. The magic of names had cast its spell upon us. So when, in a thirty-year-old gardening book, we came across a flower called *Chelidonium Japonicum*, Noel and I felt we must procure it for our new bed. We have, as yet, no idea what it will look like; it belongs, we gather, to the celandine family. Perhaps it will be a dull little plant, or it may even be ugly; perhaps that is the reason why hardly anybody stocks it, and it has gone into disrepute. One can, I think, rely on the popular judgment in flowers, even as one finds that the tourist-ridden places always are – or were – the loveliest. But perhaps the poor little flower has merely gone out of fashion, for there are fashions in the flower world as arbitrary as those in the world of clothes. People say that fifty years ago the lupin was almost unknown. I can myself remember the snapdragon when it was called snapdragon and not antirrhinum, and when it appeared in unenterprising gardens as an occasional isolated plant. While we were searching for asphodel plants a friend told us that the yellow variety, almost unprocurable these days, was once a popular favourite. I wonder what indiscretion causes the fall from grace of a flower. Who ordains that the compact, bonnet-shaped columbine of cottage gardens, a cluster of purple doves, shall be educated to grow long spurs and a variety of colours, and enter the highest social

ranks of the flower world?

We have procured our asphodels. We see the hillsides of the Mediterranean islands silver with them. We turn a bend in a mountain path in Corsica and come upon a clump of branching, tapering plants, with their veining of brown and their cloying, sickly smell; they grow out from the shelter of huge boulders, their foliage spreading over the pale, sun-drenched rocks and stones. Through our minds run romantic memories of the part they have played in mythology. We must have them in our garden, that we may travel far, even while we walk round our little half-acre plot. I have a friend who brought home irises from Delphi for the same reason; but they were dull things, she admitted. So we have imported cases of Roussillon wine for their literary association. But the *Asphodel ramosus* and the *Arum cornutum* and the *Arum dracuncula*, gathered into our garden from Mediterranean and Himalayas, will probably not be as wonderful as we expect, nor half as lovely and satisfying as the thousand ordinary mixed daffodils in the orchard beneath the apple blossom.

At last the black frost goes, leaving us still with lightless grey skies. There is a promise of rain, and we decide that we will venture to sow our grass seed. For there are many bare patches in the garden since we have worked out Noel's new plan. We flatten the earth and Noel sows the seed. It is man's work, symbolic in its action. He is happy as he flings the seed on the ground, with the generous gesture of the sower. I cover it with sifted earth, and then we roll it and lay over it disused pea sticks, that the birds shall not get at the seed. That night, as I lie in bed, I hear a long, gentle rain, and we are hoping that the seed will soon germinate.

Before we settle down to work, we promise ourselves a short walk round the garden. But it must be short and quick, for we have much to do. If one would walk quickly round the garden one should wear shackles on one's hands. It is bad enough to have hands free, but it is disastrous to find oneself armed with trowel or fork. And I have a fork and Noel has a trowel. Should we find life dull if our garden were completely cleared of weeds? I rather imagine that we should, for it is with such rapture that we dart off from the grass walks to destroy a sprouting charlock. And as Noel has a trowel with him, it is a shame not to dig up those young dandelions among the rose bushes. I wait for him as he digs up dandelions, and he waits for me as I pull groundsel from the vegetable beds, where it hides beneath the spreading stems of the purple sprouting broccoli. As we pass by the tool shed Noel remembers that there is an overhanging branch that should be removed from the willow in the back hedge; so what is easier than to fetch the saw and cut it now, before we forget it?

And I must run along and have a look at the broccoli, to see if they are ready yet for cutting. Just as I have resolved to ignore everything and merely walk, faded crocuses attract Noel's attention; I help him remove their frayed, brown bodies. And so our walk tarries. It is bad enough in March, but it is a great deal worse in May. When the sun brings dandelions to seed in the orchard within the space of an hour, half a dozen walks a day are not enough to save us from having to pause at every step to remove the dandelion clocks. I have known a "quick walk round the garden" occupy nearly an hour. We stop in our collecting of seeding dandelions only when both hands and all our pockets are full of the downy heads. Once we are in the kitchen, burning them in the stove, we may be persuaded to behave ourselves and settle down to work, but I have known Noel so waylaid by other affairs in the garden that he has forgotten to empty his pockets of dandelions seeds, and has brought out his handkerchief at a tea party a day or two later and sown dandelions over our hostess's drawing-room floor. Another dangerous enemy of "a quick walk" is the viola. There are always dead violas and pansies to remove. I remember a friend who came to stay with us and turned out to be a violent gardener; she would disappear for stretches at a time, and for a long while we did not know what happened to her: at last she was discovered on her knees beyond the long grasses in the orchard, picking off dead pansy heads. "Let me alone," she pleaded, "this makes me feel I really am doing something."

Sun bursts out from behind the clouds, opening wide the crocuses in the middle of the lawn. We think too often of flowers as colour, ignoring their form. This is especially so in the case of the crocus, whose masses of bright colour are particularly welcome after the grey of winter; but the crocus should be studied individually, that one may appreciate the extreme beauty of its shape and markings. Here, on the lawn, there are so many different kinds: pure white with orange centres, plain pale mauve, plain dark purple, white with purple at the base, white with purple stripes coming up all over it, mauve striped, mauve half-plain half-striped, redder mauve. And as the markings vary, so do the curves of the flowers themselves.

As we plant some new campanulas, I am surprised at the difference in the roots of varieties of one family. The *Platycodon* has long deep fleshy roots; some are untidy and spreading and others are thread-like. They are all very difficult to plant. But the hardest of all things to plant is *Asphodel ramosus*. Its roots hang down from the main stalk like a bunch of uniform sized, dull coloured carrots. One must make a large hole, with a pyramid of earth in the middle, and lay the several roots against and round it. Compared with this, any planting will be easy.

The warm weather is surely coming, for Lacey has removed the piece of string that he wears round his waist during the winter to prevent the wind from blowing up his back. One should not ignore little signs. Soon the air will be soft and it will be sowing weather.

As we wait for the time of the sowing of seeds, we prepare for planting our new alpines. We have a grand idea of filling in the

straight lines between the cement paving of our shelter with tiny plants. But first we must provide good earth. We look at each other guiltily, for we know that the same idea has come to both of us. We will steal some good earth. We take baskets with us, covering our trowels with harmless looking sacking, and go for a walk on the hills and in the woods. Should we meet anyone, we will pretend to look at the distant view. Should we encounter a keeper – well, let us hope that we do not. Suddenly I am a small child again, trespassing in the woods for primroses, and the keeper is an enormous giant with legs as tall as a young tree, and a head that touches the lowest clouds. I watch and listen for every movement and sound, and it takes me a very long time to fill my basket with the good earth. As we return, with brimming baskets, we notice the wonderful friable earth of the mole heaps on the top of the hill. It would be even better for our purpose than this decayed leaf mould we have stolen. We risk a second journey, and by evening the straight lines between the cement paving are full of chocolate coloured earth.

Then the parcel of new alpines arrives. Lacey opens it for us, noticing that the package has been broken in the post.

"There now," he says, "that always happens with me, too. Never a parcel of seeds or plants comes to me but what the postman hasn't opened it at one corner, to peep in and see what we are going to have growing in our garden this year. Never a time but what this happens." He is an enviable person, that he should suppose even the postman to be interested in his garden.

We plant the new alpines. As we look up their obscure names in our gardening books we are not much the wiser. Even Lacey does not pretend to know them all. We open them out, fourteen of them, and fit the round plants into the long lines and crevices, taking care never to break or damage a fibre of the delicate roots. We know the various stonecrops, the mints and pinks, but we wait now with

great excitement to see what flowers will bloom above the minute foliage of the rest of the plants. In the runnels between the plants we sow seeds of alpines. They are the smallest seeds I have ever handled and are supposed to produce tiny pinks, campanulas, sedums, and purple cress. There is magic in these microscopic seeds, spells that worked from the beginning of time, more wonderful than any blazened marvels of the modern inventor. Perhaps it is as well that we take them for granted. If we did not, we should find a garden a perpetually exhausting nervous strain.

We have an orgy of work in the garden. For three whole days we dig and bend and weed and plant, till our softened limbs ache and grow stiff and our brains cease to fret and worry. Lacey laughs at us; for, he says, he is never stiff. But our tiredness feels good, and we have at last cleared some of the worst weeds from the garden. One special bed had weighed upon our conscience for months past, and we had always managed to find a good excuse for postponing work on it. It was like a year's uncombed head of hair, blown and tangled by wind and new growth. Poppies, grass and weeds of every description submerged forget-me-nots, campanulas and tulips, giving to the bed the look of the proverbial deserted garden. At last we struggled with ourselves and vowed to clear it, and now we are happier. The mint bed, too, has been cleared; buttercups had stretched over it and dug their roots into it and choked it. We cannot keep pace in our garden with this everlasting growth of weeds.

Now warmth comes to the earth. We work in the garden in the hot sun with our sleeves turned up. Within a few minutes we have seen the first violet and the first yellow butterfly. The first thrush is building her nest in the hawthorn hedge. It is the time of year when we look about us greedily for the first sight of things. During the winter months we have been thankful for the budding and flowering of individual plants, extracting the utmost possible enjoyment out

of the few things that have bloomed. Until today we have been reconciled to dark and wet and storm, and have forgotten what a spring day could be. Chivalrously defending the winter, we have even disparaged the spring, but it turns out to be all that it should be. No extravagance is too much for it today. A few days of sun have opened the daffodils in the orchard and brought the hyacinths to their height. All around us in the garden are flowering anemones and aubretia, wallflowers, and primulas. It is as though some power has suddenly given the signal "Go!" and everything rushes as it were in a headlong charge. My precious little *Tulipa kaufmanniana* is in perfect shape and bloom, showing itself to me for the first time; and I am not disappointed. One feels the organic unity that links its pointed petals of yellow and pink with the leaves that curl tight round its stalk. In the vegetable patch, broad beans show inches high in their rows, and lettuces are thick in lines of pale green. The crown of beauty in the rush of this precocious spring belongs to two of our plum trees, white with blossom. Across the fragrant air come the sounds of sheep and lamb. Over the fence the young goat suckles two white kids. The flutey note comes into the blackbird's song. The hedges rustle with building birds.

Lacey arrives early this morning, with a soft look on his face. He is thinking of sowing seeds. Tenderly, like a shepherd with his lambs, he walks the rows, sprinkling the small seeds into the ground with his enormous hands, covering them gently with the warm earth, and weaving a pattern of black thread across the vegetable beds, that the sparrows may not eat them. Throughout the day he sows parsnip and carrot, leek and pea. The year has begun.

Please contact Little Toller Books
to join our mailing list or for more information
on current and forthcoming titles.

Nature Classics Library

THE JOURNAL OF A DISAPPOINTED MAN *W.N.P. Barbellion*
THROUGH THE WOODS *H.E. Bates*
MEN AND THE FIELDS *Adrian Bell*
ISLAND YEARS, ISLAND FARM *Frank Fraser Darling*
A SHEPHERD'S LIFE *W.H. Hudson*
WILD LIFE IN A SOUTHERN COUNTY *Richard Jefferies*
FOUR HEDGES *Clare Leighton*
LETTERS FROM SKOKHOLM *R.M. Lockley*
THE UNOFFICIAL COUNTRYSIDE *Richard Mabey*
RING OF BRIGHT WATER *Gavin Maxwell*
THE SOUTH COUNTRY *Edward Thomas*
SALAR THE SALMON *Henry Williamson*

Also Available

THE LOCAL *Edward Ardizzone & Maurice Gorham*
A long out-of-print celebration of London's pubs
by one of Britain's most-loved illustrators.

LITTLE TOLLER BOOKS
Stanbridge Wimborne Minster Dorset BH21 4JD
Telephone: 01258 840549
ltb@dovecotepress.com
www.dovecotepress.com